SpringerBriefs in Physics

SpringerBriefs in Physics are a series of slim high-quality publications encompassing the entire spectrum of physics. Manuscripts for SpringerBriefs in Physics will be evaluated by Springer and by members of the Editorial Board. Proposals and other communication should be sent to your Publishing Editors at Springer.

Featuring compact volumes of 50 to 125 pages (approximately 20,000–45,000 words), Briefs are shorter than a conventional book but longer than a journal article. Thus, Briefs serve as timely, concise tools for students, researchers, and professionals.

Typical texts for publication might include:

- A snapshot review of the current state of a hot or emerging field
- A concise introduction to core concepts that students must understand in order to make independent contributions
- An extended research report giving more details and discussion than is possible in a conventional journal article
- A manual describing underlying principles and best practices for an experimental technique
- An essay exploring new ideas within physics, related philosophical issues, or broader topics such as science and society

Briefs allow authors to present their ideas and readers to absorb them with minimal time investment. Briefs will be published as part of Springer's eBook collection, with millions of users worldwide. In addition, they will be available, just like other books, for individual print and electronic purchase. Briefs are characterized by fast, global electronic dissemination, straightforward publishing agreements, easy-to-use manuscript preparation and formatting guidelines, and expedited production schedules. We aim for publication 8–12 weeks after acceptance.

More information about this series at http://www.springer.com/series/8902

Yasuhiro Asano

Andreev Reflection in Superconducting Junctions

Springer

Yasuhiro Asano
Department of Applied Physics
Hokkaido University
Sapporo, Hokkaido, Japan

ISSN 2191-5423 ISSN 2191-5431 (electronic)
SpringerBriefs in Physics
ISBN 978-981-16-4164-0 ISBN 978-981-16-4165-7 (eBook)
https://doi.org/10.1007/978-981-16-4165-7

This Springer imprint is published by the registered company Springer Nature Singapore Pte Ltd.
The registered company address is: 152 Beach Road, #21-01/04 Gateway East, Singapore 189721, Singapore

Contents

Acronyms

BCS	Bardeen-Cooper-Schrieffer
BdG	Bogoliubov-de Gennes
CPR	Current-phase relationship
NS	Normal-metal/superconductor
SIS	Superconductor/insulator/superconductor
SNS	Superconductor/normal-metal/superconductor
ZES	Zero-energy state

Chapter 1
Introduction

Abstract On the basis of mean-field theory of superconductivity, we discuss low-energy transport properties in superconducting junctions in this book. Andreev reflection at a junction interface is a key phenomenon to understand characteristic feature of transport properties in superconducting heterostructure. The purpose of this chapter is to make clear the focus of this book. We also overview superconducting phenomena briefly to imply an importance of quantum mechanics. Particular attention should be paid to roles of the phase of superconducting condensate in the flux quantization and the Josephson effect.

1.1 Scope of This Book

Superconductivity is a macroscopic quantum phenomenon discovered by H. K. Onnes in 1911. He and his colleague observed the disappearance of electric resistivity in Hg at 4.2 Kelvin (K) (Onnes 1911; van Delft and Kes 2010). This experimental finding opened a new era of low-temperature science. Superconductors exhibit various unusual responses to an electromagnetic field such as perfect diamagnetism, persistent current, and flux quantization. In 1957, Bardeen-Cooper-Schrieffer (BCS) explained the microscopic mechanism of superconductivity (Bardeen et al. 1957). According to BCS theory, two electrons on the Fermi surface form a pair due to an attractive interaction mediated by phonons at a low temperature. The electron pair is called Cooper pair. The coherent condensation of a number of Cooper pairs causes the unusual response to electromagnetic fields. The mean-field theory by Bogoliubov enabled us to describe the thermodynamic properties of a superconductor as well as the linear response to electromagnetic fields. Before publishing the BCS theory, however, a phenomenological theory by the London brothers and that by Ginzburg-Landau explained the electromagnetic properties of superconductors successfully. The assumptions introduced in these phenomenological theories have been justified by the microscopic theory of superconductivity.

It is not easy to explain all of the exotic superconducting phenomena within a short textbook. Fortunately, historical textbooks have achieved such a goal successfully (Tinkham 1996; Abrikosov 1988; Nakajima 1971). In this textbook, we focus

Y. Asano, *Andreev Reflection in Superconducting Junctions*,
SpringerBriefs in Physics, https://doi.org/10.1007/978-981-16-4165-7_1

only on the electric transport properties in superconducting heterostructures such as the differential conductance in a normal-metal/superconductor (NS) junction and the Josephson current in a superconductor/insulator/superconductor (SIS) junction. Generally speaking, electric current is obtained theoretically by calculating Green's function of a junction. In this textbook, however, we do not use such technique at all. The reason is very simple. Green's function is undoubtedly a powerful tool for experts but is not always the useful language for general researchers in solid-state physics. The goal of this book is to demonstrate exotic transport properties of a junction consisting of an unconventional superconductor. The contents in the following sections are written to be straight to the goal. We begin with superconducting phenomena observed in experiments to make clear theoretical requirements to explain these phenomena. Chapters 2 and 3 are spent relating to theoretical tools and basics of superconductivity, which are the minimum necessary items to understand superconductivity. Therefore, readers should study more details on BCS, Ginzburg-Landau, and London theories in standard textbooks (de Gennes 1966; Mahan 1990; Schrieffer 1988; Abrikosov et al. 1975; Fetter and Walecka 1971; Nakajima 1971). Chapters 4–6 are the main body of the textbook, where the transport properties of superconducting junctions are explained in terms of the Andreev reflection. The outline of the theoretical method in these chapters is as follows. At first, the wave functions are obtained by solving Bogoliubov-de Gennes (BdG) equation near a junction interface. Secondly, the transport coefficients are calculated by connecting the wave functions under appropriate boundary conditions at the junction interface. Finally, the substitution of the resulting coefficients into the current formulas enables us to analyze transport properties measurable in experiments. Instead of deriving the current formulas, we explain the physical picture that relates the Andreev reflections to the transport phenomena. In Chap. 7, we briefly discuss the anomalous proximity effect in a dirty normal metal attached to an unconventional superconductor.

1.2 Overview of Superconducting Phenomena

Let us begin a brief overview of superconducting phenomena with the discovery of superconductivity in 1911. Low temperature physics might be a trend in science at the beginning of the 20th century. The last inactive gas He became liquid at 4.2 K in 1908. A natural question would be what happens on materials near zero temperature. The resistance R in a metal was an issue in this direction. It was accepted at that time that a charged particle (electron) moves freely in a metal and carries the electric current I under the bias voltage V. Ohm's law $V = IR$ holds true in metals for wide temperature range. According to kinetics of gases by Boltzmann, all particle stop their motion at zero temperature. Then, a metal should be an insulator at zero temperature because all electrons stop moving. The resistivity of a metal would be infinitely large at zero temperature. On the other hand, there was a group of scientists who speculated, based on the experimental results at that time, that the resistivity would be zero at zero temperature. An electron is scattered by the thermal vibration of ions in the

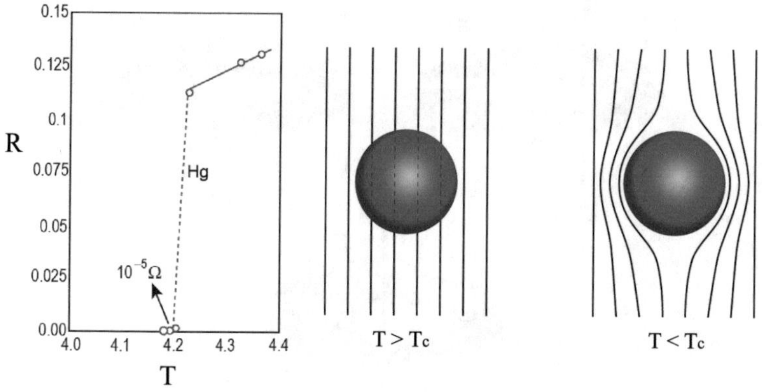

Fig. 1.1 (Left) An image of the historic plot (Onnes 1911; van Delft and Kes 2010). The resistance in Hg vanishes at 4.2 K. (Right) A superconductor excludes magnetic flux at a temperature lower than T_c and a field below a certain critical one

metal, which is the cause of the finite resistance. At zero temperature, therefore, an electron under bias voltage would move freely at zero temperature. The experiment by Onnes was supposed to settle the debate over. The experimental data on Hg, however, showed the disappearance of the resistivity at a temperature above zero as shown in Fig. 1.1. This was the first quantum phenomenon that we observed in our history. The phenomenon was named superconductivity. The temperature that divides the two transport phases is called transition temperature or critical temperature T_c. A series of experiments reported anomalies in thermodynamic property of a superconductor at T_c. The specific heat shows a jump at T_c, which suggested the rapid decrease of the entropy and the development of an order below T_c.

The following is a brief summary of the peculiar response of superconductors to magnetic fields. The first phenomenon is called Meissner effect. An external magnetic field $\boldsymbol{H}$ threats a normal metal at a room temperature. Far below the transition temperature $T \ll T_c$, the metal excludes a weak external magnetic field from its interior as shown in Fig. 1.1. The magnetic flux density B in a superconductor is zero as represented by

$$\boldsymbol{B} = \boldsymbol{H} + 4\pi \boldsymbol{M} = 0, \tag{1.1}$$

where $\boldsymbol{M}$ denotes the magnetization of a superconductor. The magnetic susceptibility defined by $\boldsymbol{M} = \chi \boldsymbol{H}$ results in

$$\chi = -\frac{1}{4\pi}. \tag{1.2}$$

Superconductors are fully diamagnetic at temperatures much lower than T_c in weak magnetic fields.

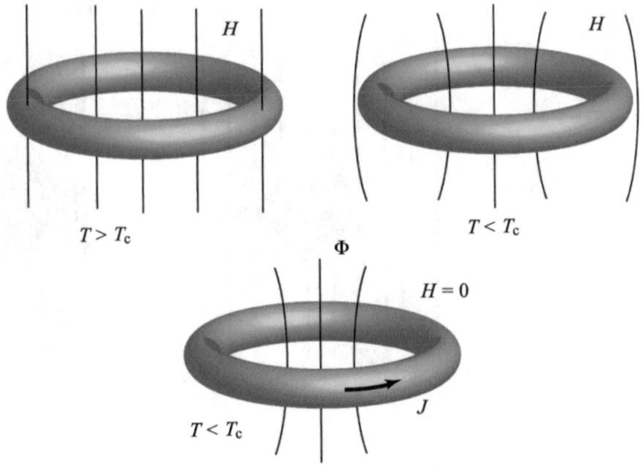

Fig. 1.2 Persistent current at a ring-shaped superconductor. At a higher temperature than T_c, a uniform external magnetic field is applied as shown in the upper left figure. As shown in the upper right figure, the superconductor is cooled to a temperature much lower than T_c while maintaining the magnetic field. The magnetic field is excluded from the superconductor due to the Meissner effect. Then, the external magnetic field is switched off as shown in the lower figure. A quantized magnetic flux Φ remains in the ring. The current persists as long as $T \ll T_c$

The magnetic response of a superconductor with multiply-connected geometry is more unusual. We apply a magnetic field to a ring at $T > T_c$ as shown in Fig. 1.2, where we illustrate the cooling process under a magnetic field. When a temperature is decreased down below T_c, a ring-shaped superconductor excludes the magnetic fluxes from its arm. After switching off a magnetic field at $T \ll T_c$, the magnetic flux is trapped in the hole of the ring. According to the Maxwell equation

$$\nabla \times \boldsymbol{H} = \frac{4\pi}{c} \boldsymbol{j}, \tag{1.3}$$

the magnetic flux in the hole generates the electric current that circulates through the ring. The current is called persistent current because it does not decrease at all even after a long time. The results suggest that superconducting current flows in equilibrium as long as the magnetic flux is penetrating the hole in the ring. A similar phenomenon is known as orbital diamagnetism $\boldsymbol{M} = \chi \boldsymbol{H}$ in metals. An external magnetic field generates the electric current in equilibrium as shown in Fig. 1.3. The susceptibility χ is negative in a metal. Together with Eq. (1.3), we find

$$\boldsymbol{j} = \frac{c}{4\pi\chi} \nabla \times \boldsymbol{M}. \tag{1.4}$$

The electric current is represented as a rotation of a vector. The total transport current can be calculated as

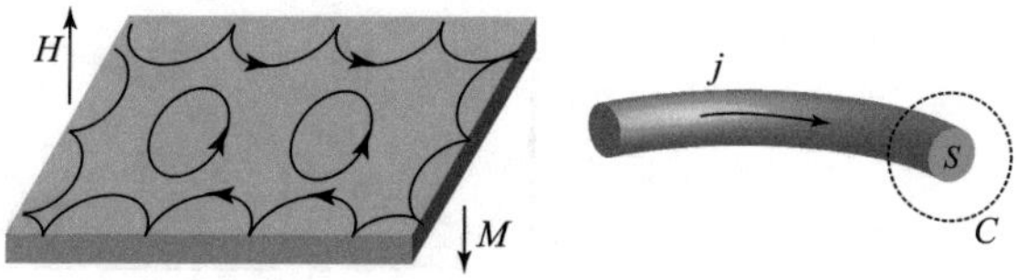

Fig. 1.3 Schematic picture of the diamagnetic current in a metal. The diamagnetic current in a metal cannot flow out from the metal

$$J = \int d\boldsymbol{S} \cdot \boldsymbol{j} = \frac{c}{4\pi\chi} \int_S d\boldsymbol{S} \cdot \nabla \times \boldsymbol{M} = \frac{c}{4\pi\chi} \oint_C d\boldsymbol{l} \cdot \boldsymbol{M}. \tag{1.5}$$

The net current is zero because the integration path C can be taken outside the metal. The magnetization is absent there, (i.e., $\boldsymbol{M} = 0$). Namely, the current described as $\boldsymbol{j} \propto \nabla \times \boldsymbol{M}$ does not flow out from the metal. As we will see later, the supercurrent can be represented as $\boldsymbol{j} \propto \boldsymbol{A}$ and flows out from a superconductor. In addition to the persistence of the current in a superconducting ring, the magnetic flux trapped in the hole can be quantized as

$$\Phi = n\phi_0, \quad \phi_0 = \frac{\pi\hbar c}{e}, \tag{1.6}$$

where n is an integer number. In quantum mechanics, the quantization of physical values is a consequence of the boundary conditions of the wave function. As we will see later, the macroscopic wave function of Cooper pairs must be single-valued in real space, which explains the flux quantization.

The perfect screening of a magnetic field is realized when the magnetic field is weak enough. At higher magnetic fields, the magnetic response depends on the types of superconductors. The type I superconductor excludes magnetic fields up to the critical magnetic field H_c as shown in Fig. 1.4a. All superconducting properties disappear for $H > H_c$. The magnetic properties of type II superconductor are characterized by two critical fields: H_{C_1} and H_{C_2} as shown in Fig. 1.4b. Such a type II superconductor excludes magnetic field for $H < H_{C_1}$ and becomes a normal metal for $H > H_{C_2}$. In an intermediate magnetic field for $H_{C_1} < H < H_{C_2}$, the metal remains superconducting, but magnetic flux enters the metal as a number of quantized vortices. The magnetic flux is quantized as ϕ_0 at each vortex. As a result, the profile of the magnetic flux is non-uniform in real space. Near H_{C_2}, a number of vortex form a hexagonal structure called Abrikosov lattice. Ginzburg-Landau showed that the boundary of the two types of superconductors is described by a parameter $\kappa = \lambda_L/\xi_0$, where ξ_0 is called coherence length and λ_L is London's penetration length. The meaning of the two length scales will be explained in Chap. 3. The type I (type II) superconductors are characterized by $\kappa < 1/\sqrt{2}$ ($\kappa > 1/\sqrt{2}$).

In quantum mechanics, we know that the wave function is a complex number

$$\psi(\boldsymbol{r}) = |\psi(\boldsymbol{r})|e^{i\varphi}. \tag{1.7}$$

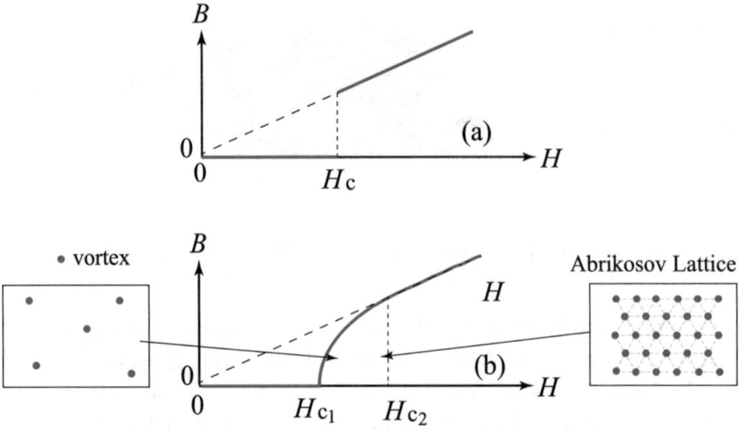

Fig. 1.4 $B - H$ curve in a superconductor: **a** type I and **b** type II

The amplitude $|\psi(\boldsymbol{r})|^2$ is proportional to the probability of observing a particle there. A uniform phase in real space φ plays no role in the Schrödinger picture, which describes the quantum state of a single particle. As we will see later, the superconducting state is described phenomenologically well by the macroscopic wave function,

$$\psi(\boldsymbol{r}) = \sqrt{n(\boldsymbol{r})}\, e^{i\varphi}. \tag{1.8}$$

Here, $\psi(\boldsymbol{r})$ describes a quantum state consisting of N_{A} Cooper pairs with $N_{\mathrm{A}} \sim 10^{23}$ being the Avogadro number. Namely, N_{A} Cooper pairs share the common phase φ and form a phase-coherent condensate. This causes the anomalous response of a superconductor to the electromagnetic field. The Josephson effect is a highlight of superconducting phenomena (Josephson 1962). The electric current flows between the two superconductors in equilibrium. Figure 1.5 shows a typical Josephson junction where an insulating film is sandwiched by two superconductors. Today we know that the current can be represented as

$$J = J_c \sin(\varphi_L - \varphi_R), \tag{1.9}$$

where J_c is called critical current and $\varphi_{L(R)}$ is called superconducting *phase* at the left (right) superconductor.

What is the superconducting phase? Why does the phase difference generate the electric current? These are essential questions that we would like to answer. As a first step in this direction, let us consider Feynman's discussion of the Josephson effect. He hypothesized that a superconducting condensate containing N_{A} electrons behaves as if it were a coherent object. This view enables us to introduce the wave function of the superconducting condensate

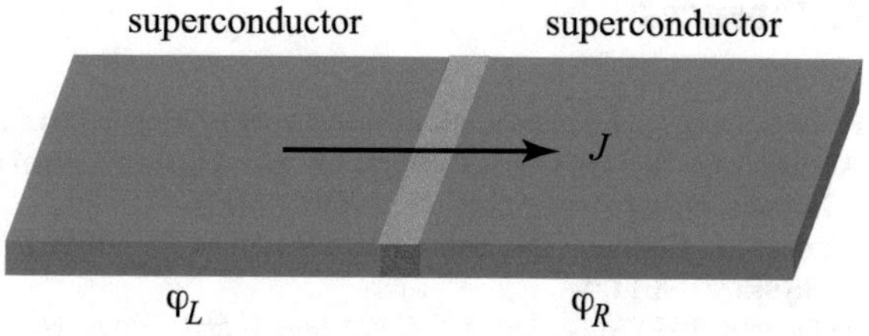

Fig. 1.5 Two superconductors are separated by a thin insulating barrier. The electric current flows through the junction in the presence of the phase difference across the junction

$$\psi_L = \sqrt{n_L} e^{i\varphi_L}, \quad \psi_R = \sqrt{n_R} e^{i\varphi_R}, \tag{1.10}$$

where $n_{L(R)}$ is the electron density in the superconducting condensate at the left (right) superconductor and $\varphi_{L(R)}$ is the phase of the wave function. The two superconducting condensates interact with each other through a thin insulating barrier. Let us introduce the tunneling Hamiltonian between the two superconductors. The rate equation would be given by

$$i\hbar \frac{\partial}{\partial t} \psi_L = T_I \, \psi_R, \quad i\hbar \frac{\partial}{\partial t} \psi_R = T_I \, \psi_L \tag{1.11}$$

where T_I is the tunneling element of the insulating barrier. The electric current density through the barrier is given by

$$j = -e \frac{\partial}{\partial t} n_L, \quad n_L = \psi_L^* \, \psi_L. \tag{1.12}$$

The time derivative of density is calculated as

$$\frac{\partial}{\partial t} n_L = \partial_t \psi_L^* \, \psi_L + \psi_L^* \, \partial_t \psi_L = \frac{i}{\hbar} T_I \, \psi_R^* \, \psi_L - \frac{i}{\hbar} T_I \, \psi_L^* \, \psi_R, \tag{1.13}$$

$$= \frac{i}{\hbar} T_I \sqrt{n_L n_R} \left(e^{i(\varphi_L - \varphi_R)} - e^{-i(\varphi_L - \varphi_R)} \right), \tag{1.14}$$

$$= -2 \frac{T_I}{\hbar} \sqrt{n_L n_R} \sin(\varphi_L - \varphi_R). \tag{1.15}$$

Then, the electric current results in

$$J = \frac{2e}{\hbar} T_I \, S \sqrt{n_L n_R} \sin(\varphi_L - \varphi_R), \tag{1.16}$$

where S is the cross section of the junction. This equation explains well the Josephson effect. Feynman's argument clearly illustrates the essence of superconductivity. His assumption in Eq. (1.10) suggests that the phase of the wave function characterizes the ground state of many electrons. However, it would not be easy for us to imagine such a ground state for electrons in a metal.

References

Abrikosov, A.A.: Fundamentals of the Theory of Metals. North-Holland, Amsterdam (1988)
Abrikosov, A.A., Gor'kov, L.P., Dzyaloshinski, I.E.: Methods of Quantum Field Theory in Statistical Physics. Dover Publications, New York (1975)
Bardeen, J., Cooper, L.N., Schrieffer, J.R.: Phys. Rev. **108**, 1175 (1957). https://doi.org/10.1103/PhysRev.108.1175
de Gennes, P.G.: Superconductivity of Metals and Alloys. Benjamin, New York (1966)
Fetter, A.L., Walecka, J.D.: Quantum Theory of Many-Particle Systems. McGraw-Hill, New York (1971)
Josephson, B.: Phys. Lett. **1**(7), 251 (1962). https://doi.org/10.1016/0031-9163(62)91369-0
Mahan, G.D.: Many-Particle Physics. Plenum Press, New York (1990)
Nakajima, S.: Introduction to Superconductivity (Japanese). Baifukan, Tokyo (1971)
Onnes, H.K.: Commun. Phys. Lab. Univ. Leiden Suppl. 29 (1911)
Schrieffer, J.R.: Theory of Superconductivity. Addison-Wesley, New York (1988)
Tinkham, M.: Introduction to Superconductivity. McGraw-Hill, New York (1996)
van Delft, D., Kes, P.: Phys. Today **63**, 38 (2010). https://doi.org/10.1103/10.1063/1.3490499

Chapter 2
Theoretical Tools

Abstract Today we know that superconductivity is a many-body effect of electrons in a metal and that the attractive interaction between two electrons plays an important role in stabilizing the superconducting state. In order to describe superconducting phenomena, several theoretical tools must be used. The purpose of this chapter is to explain what tools we need and how to use them. The first tool is the creation and the annihilation operators of a particle. After explaining how to calculate quantum and thermal average of operator, we will discuss properties of the coherent state of Bosons and its relation to the superconducting state.

2.1 Harmonic Oscillators

The Hamiltonian of a harmonic oscillator in one-dimension reads

$$H = \frac{p_x^2}{2m} + \frac{1}{2}m\omega^2 x^2, \quad p_x = \frac{\hbar}{i}\frac{d}{dx}, \tag{2.1}$$

in the Schrödinger picture. The energy eigenvalues are given by

$$E = \hbar\omega\left(n + \frac{1}{2}\right), \quad n = 0, 1, 2, \cdots. \tag{2.2}$$

Let us introduce a length scale x_0 which balances the kinetic energy and the potential energy as

$$\frac{\hbar^2}{2mx_0^2} = \frac{1}{2}m\omega^2 x_0^2. \tag{2.3}$$

The solutions $x_0 = \sqrt{\hbar/m\omega}$ is the characteristic length scale of the Hamiltonian. By applying the scale transformation $x = x_0 q$, the Hamiltonian and the commutation relation becomes

Y. Asano, *Andreev Reflection in Superconducting Junctions*,
SpringerBriefs in Physics, https://doi.org/10.1007/978-981-16-4165-7_2

$$H = -\frac{\hbar^2}{2mx_0^2}\frac{d^2}{dq^2} + \frac{1}{2}m\omega^2 x_0^2 q^2 = \frac{\hbar\omega}{2}\left(-\frac{d^2}{dq^2} + q^2\right), \tag{2.4}$$

$$[q, p] = i, \quad p = -i\frac{d}{dq}. \tag{2.5}$$

We define two operators as

$$a = \frac{1}{\sqrt{2}}(q + ip) = \frac{1}{\sqrt{2}}\left(q + \frac{d}{dq}\right), \quad a^\dagger = \frac{1}{\sqrt{2}}(q - ip) = \frac{1}{\sqrt{2}}\left(q - \frac{d}{dq}\right). \tag{2.6}$$

They satisfy the commutation relations

$$[a, a^\dagger] = \frac{1}{2}\left[(q + ip)(q - ip) - (q - ip)(q + ip)\right], \tag{2.7}$$

$$= \frac{1}{2}\left[q^2 - iqp + ipq + p^2 - q^2 - iqp + ipq - p^2\right] = -i[q, p] = 1, \tag{2.8}$$

$$[a, a] = 0. \tag{2.9}$$

The original operators are represented in terms of a and $a^\dagger$ by

$$q = \frac{1}{\sqrt{2}}(a + a^\dagger), \quad p = \frac{1}{\sqrt{2}i}(a - a^\dagger). \tag{2.10}$$

Using the commutation relation, it is possible to derive

$$p^2 + q^2 = \frac{1}{2}\left[-(a - a^\dagger)(a - a^\dagger) + (a + a^\dagger)(a + a^\dagger)\right], \tag{2.11}$$

$$= \frac{1}{2}\left[-a^2 + aa^\dagger + a^\dagger a - (a^\dagger)^2 + a^2 + aa^\dagger + a^\dagger a + (a^\dagger)^2\right], \tag{2.12}$$

$$= 2a^\dagger a + 1. \tag{2.13}$$

Thus, the Hamiltonian is diagonalized as

$$\mathcal{H} = \hbar\omega\left(a^\dagger a + \frac{1}{2}\right). \tag{2.14}$$

The expectation value should be $\langle a^\dagger a\rangle = n$ so that the expectation value of $\mathcal{H}$ coincides with Eq. (2.2). The operator $\hat{n} = a^\dagger a$ is called number operator and satisfies the relations

$$[\hat{n}, a] = a^\dagger aa - aa^\dagger a = a^\dagger aa - (1 + a^\dagger a)a = -a, \tag{2.15}$$

$$[\hat{n}, a^\dagger] = a^\dagger aa^\dagger - a^\dagger a^\dagger a = a^\dagger(1 + a^\dagger a) - a^\dagger a^\dagger a = a^\dagger. \tag{2.16}$$

Let us assume that $|\nu\rangle$ is the eigenstate of $\hat{n}$ belonging to the eigenvalue ν,

$$\hat{n}|\nu\rangle = \nu|\nu\rangle, \quad \langle\nu'|\nu\rangle = \delta_{\nu,\nu'}. \tag{2.17}$$

It is easy to confirm the relations

$$\hat{n}a|\nu\rangle = a^\dagger a a|\nu\rangle = (-1 + aa^\dagger)a|\nu\rangle = (-a + a\nu)|\nu\rangle = (\nu - 1)a|\nu\rangle, \tag{2.18}$$
$$\hat{n}a^\dagger|\nu\rangle = a^\dagger a a^\dagger|\nu\rangle = (a^\dagger + a^\dagger\hat{n})|\nu\rangle = (a^\dagger + a^\dagger\nu)|\nu\rangle = (\nu + 1)a^\dagger|\nu\rangle. \tag{2.19}$$

The results suggest that $a|\nu\rangle$ is the eigenstate of $\hat{n}$ belonging to $\nu - 1$ and that $a^\dagger|\nu\rangle$ is the eigenstate of $\hat{n}$ belonging to $\nu + 1$. Therefore, it is possible to describe these states as

$$a|\nu\rangle = \beta|\nu - 1\rangle, \quad a^\dagger|\nu\rangle = \gamma|\nu + 1\rangle, \tag{2.20}$$

with β and γ being c-numbers. Together with the Hermite conjugate of them

$$\langle\nu|a^\dagger = \beta^*\langle\nu - 1|, \quad \langle\nu|a = \gamma^*\langle\nu + 1|, \tag{2.21}$$

the norm of these states are calculated as

$$\langle\nu|a^\dagger a|\nu\rangle = |\beta|^2, \quad \langle\nu|aa^\dagger|\nu\rangle = |\gamma|^2. \tag{2.22}$$

Simultaneously, the left-hand side of these equations can be represented by

$$\langle\nu|a^\dagger a|\nu\rangle = \langle\nu|\hat{n}|\nu\rangle = \nu, \quad \langle\nu|aa^\dagger|\nu\rangle = \langle\nu|\hat{n} + 1|\nu\rangle = \nu + 1, \tag{2.23}$$

because of the definition in Eq. (2.17). Thus, we find $\beta = \sqrt{\nu}$, $\gamma = \sqrt{\nu + 1}$, and

$$a|\nu\rangle = \sqrt{\nu}|\nu - 1\rangle, \quad a^\dagger|\nu\rangle = \sqrt{\nu + 1}|\nu + 1\rangle. \tag{2.24}$$

The relation $\nu \geq 0$ must be true because ν is the norm of $a|\nu\rangle$ as shown in Eq. (2.23). The recursive relation Eq. (2.24)

$$a|\nu\rangle = \sqrt{\nu}|\nu - 1\rangle, \tag{2.25}$$
$$a^2|\nu\rangle = \sqrt{\nu(\nu - 1)}|\nu - 2\rangle, \tag{2.26}$$
$$a^M|\nu\rangle = \sqrt{\nu(\nu - 1)\cdots(\nu - M + 1)}|\nu - M\rangle, \tag{2.27}$$

however, suggests that $\nu - M$ can be negative if we can chose large enough M. To avoid such unphysical situation, we impose the boundary condition of the series of the recursive relation. The eigenvalue of $\hat{n}$ include zero so that

$$a|0\rangle = 0 \tag{2.28}$$

deletes the physical state. The relation in Eq. (2.28) defines the vacuum of a particle $|0\rangle$. Finally we obtain

$$a|n\rangle = \sqrt{n}|n-1\rangle, \quad a^\dagger|n\rangle = \sqrt{n+1}|n+1\rangle, \quad \hat{n}|n\rangle = n|n\rangle. \tag{2.29}$$

These relations represent the mechanics of harmonic oscillators. The operator a ($a^\dagger$) is called annihilation (creation) operator because it decreases (increases) the number of oscillator by one. The ground state is then given by $|0\rangle$ where no harmonic oscillator exists. The mth excited state $|m\rangle$ describes the state in which m oscillators are excited on the vacuum as

$$|m\rangle = \frac{(a^\dagger)^m}{\sqrt{m!}}|0\rangle. \tag{2.30}$$

Next we discuss the wave function in the Schrödinger picture which can be obtained by multiplying $\langle q|$ to the relation $a|0\rangle = 0$. The representation becomes

$$0 = \langle q|a|0\rangle = \frac{1}{\sqrt{2}}\langle q|q + ip|0\rangle = \frac{q}{\sqrt{2}}\langle q|0\rangle + \frac{1}{\sqrt{2}}\frac{d}{dq}\langle q|0\rangle. \tag{2.31}$$

We obtained the differential equation

$$\frac{d}{dq}\langle q|0\rangle = -q\langle q|0\rangle. \tag{2.32}$$

The solution is given by

$$\langle q|0\rangle = C\, e^{-\frac{1}{2}q^2}. \tag{2.33}$$

The left-hand side is the definition of the wave function at the ground state and the right-hand side is Gauss's function. The wave function at the excited states are calculated as

$$\langle q|1\rangle = \langle q|a^\dagger|0\rangle = \frac{1}{\sqrt{2}}\left(q - \frac{d}{dq}\right) C\, e^{-\frac{1}{2}q^2}, \tag{2.34}$$

$$\langle q|n\rangle = \frac{1}{\sqrt{n!}}\langle q|(a^\dagger)^n|0\rangle = \frac{1}{\sqrt{n!}\sqrt{2}^n}\left(q - \frac{d}{dq}\right)^n C\, e^{-\frac{1}{2}q^2}. \tag{2.35}$$

This sequence generates the series of Hermite's polynomial.

The observable values are calculated as the average of operators. Here, the average has double meanings: the average in quantum mechanics and the average in statistical mechanics. The average of an operator $\mathcal{Q}$ is defined by

$$\langle \mathcal{Q}\rangle = \frac{1}{\Xi}\mathrm{Tr}\left[e^{-\beta\mathcal{H}}\mathcal{Q}\right], \quad \Xi = \mathrm{Tr}\left[e^{-\beta\mathcal{H}}\right], \tag{2.36}$$

where $\mathcal{H}$ is the Hamiltonian under consideration and $\beta = 1/k_B T$ corresponds to the inverse of temperature with k_B being the Boltzmann constant. To proceed the calculation, we define Tr in the average and explain several identities for the trace of matrices. Let us assume that $|p\rangle$ is the eigenstate of $\mathcal{H}$ belonging to the eigenvalue of ϵ_p,

$$\mathcal{H}|p\rangle = \epsilon_p|p\rangle, \quad \langle p'|p\rangle = \delta_{p,p'}, \quad \sum_p |p\rangle\langle p| = 1. \tag{2.37}$$

When we choose $|p\rangle$ as the basis of the trace, the partition function can be calculated as

$$\Xi = \mathrm{Tr}\left[e^{-\beta\mathcal{H}}\right] = \sum_p \langle p|e^{-\beta\mathcal{H}}|p\rangle = \sum_p e^{-\beta\epsilon_p}. \tag{2.38}$$

The first theorem says that a value of the trace does not change under the unitary transformation,

$$|q\rangle = \sum_p A_{q,p}|p\rangle, \quad \sum_p A_{q,p}A^*_{q',p} = \sum_p A^*_{p,q'}A_{p,q} = \delta_{q,q'}. \tag{2.39}$$

The second equation is the definition of the unitary matrix of A, (i.e., $A^\dagger A = A^\dagger A = 1$). The trace of $\mathcal{Q}$ is transformed as

$$\sum_p \langle p|\mathcal{Q}|p\rangle = \sum_p \sum_{q,q'} \langle q'|A^*_{p,q'}\mathcal{Q}A_{p,q}|q\rangle = \sum_{q,q'}\left[\sum_p A^*_{p,q'}A_{p,q}\right]\langle q'|\mathcal{Q}|q\rangle, \tag{2.40}$$

$$= \sum_q \langle q|\mathcal{Q}|q\rangle. \tag{2.41}$$

The second theorem says a value of the trace does not change under the cyclic permutation of operators

$$\mathrm{Tr}[A\,B\,C] = \mathrm{Tr}[C\,A\,B] = \mathrm{Tr}[B\,C\,A]. \tag{2.42}$$

It is easy to confirm the relations

$$\mathrm{Tr}[A\,B] = \sum_{p,q} A_{p,q}B_{q,p} = \sum_{p,q} B_{q,p}A_{p,q} = \mathrm{Tr}[B\,A], \tag{2.43}$$

$$\mathrm{Tr}[(A\,B)\,C] = \mathrm{Tr}[C\,(A\,B)]. \tag{2.44}$$

By using these theorems, we try to calculate the average of number operator in the Hamiltonian

$$\mathcal{H} = \sum_p \epsilon_p \, a_p^\dagger \, a_p, \quad [a_p, a_{p'}^\dagger] = \delta_{p,p'}, \quad [a_p, a_{p'}] = 0, \tag{2.45}$$

where p indicates a quantum state. The average of the number operator is calculated by

$$\langle a_p^\dagger a_q \rangle = \frac{1}{\Xi}\mathrm{Tr}[e^{-\beta\mathcal{H}} a_p^\dagger a_q] = \frac{1}{\Xi}\mathrm{Tr}[a_q e^{-\beta\mathcal{H}} a_p^\dagger], \tag{2.46}$$

$$= \frac{1}{\Xi}\mathrm{Tr}[e^{-\beta\mathcal{H}} \left(e^{\beta\mathcal{H}} a_q e^{-\beta\mathcal{H}}\right) a_p^\dagger]. \tag{2.47}$$

The Heisenberg operator is represented as

$$I(\beta) = e^{\beta\mathcal{H}} a_q e^{-\beta\mathcal{H}}, \quad \partial_\beta I = e^{\beta\mathcal{H}} [\mathcal{H}, a_q] e^{-\beta\mathcal{H}}. \tag{2.48}$$

We find $[\mathcal{H}, a_q] = -\epsilon_q a_q$ because of the relation

$$\mathcal{H}\, a_q = \sum_p \epsilon_p \, a_p^\dagger \, a_p \, a_q = \sum_p \epsilon_p (-\delta_{p,q} + a_q \, a_p^\dagger) \, a_p = a_q \mathcal{H} - \epsilon_q a_q. \tag{2.49}$$

By substituting the results into Eq. (2.48), we obtain the differential equation

$$\partial_\beta I(\beta) = -\epsilon_q I(\beta). \tag{2.50}$$

The solution is given by

$$I(\beta) = a_q e^{-\epsilon_q \beta}. \tag{2.51}$$

The average can be represented as

$$\langle a_p^\dagger a_q \rangle = \frac{1}{\Xi}\mathrm{Tr}[e^{-\beta\mathcal{H}} \, a_p^\dagger \, a_q] = \frac{1}{\Xi}\mathrm{Tr}[e^{-\beta\mathcal{H}} \, e^{\beta\mathcal{H}} \, a_q \, e^{-\beta\mathcal{H}} \, a_p^\dagger], \tag{2.52}$$

$$= \frac{1}{\Xi}\mathrm{Tr}[e^{-\beta\mathcal{H}} \, a_q a_p^\dagger] e^{-\epsilon_q \beta} = \frac{1}{\Xi}\mathrm{Tr}[e^{-\beta\mathcal{H}} (\delta_{p,q} + a_p^\dagger a_q)] e^{-\epsilon_q \beta}, \tag{2.53}$$

$$= \langle a_p^\dagger a_q \rangle e^{-\epsilon_q \beta} + \delta_{p,q} e^{-\epsilon_q \beta}. \tag{2.54}$$

We reach the final results of average,

$$\langle a_p^\dagger a_q \rangle = \delta_{p,q} \, n_B(\epsilon_q), \tag{2.55}$$

$$n_B(\epsilon_q) = \frac{1}{e^{\epsilon_q \beta} - 1} = \frac{1}{2}\left[\coth\left(\frac{\epsilon_p}{2k_B T}\right) - 1\right], \tag{2.56}$$

where n_B is the Bose-Einstein distribution function. The results are consistent with a fact that Eq. (2.45) describes the commutation relation of bosons.

2.2 Free Electrons in a Metal

The Hamiltonian for free electrons in a metal is given by

$$\mathcal{H}_0 = \sum_{\alpha=\uparrow,\downarrow} \int d\boldsymbol{r}\, \psi_\alpha^\dagger(\boldsymbol{r}) \left[-\frac{\hbar^2\nabla^2}{2m} - \epsilon_F \right] \psi_\alpha(\boldsymbol{r}), \tag{2.57}$$

$$\left\{\psi_\alpha(\boldsymbol{r}), \psi_\beta^\dagger(\boldsymbol{r}')\right\} = \delta(\boldsymbol{r}-\boldsymbol{r}')\delta_{\alpha,\beta}, \quad \{\psi_\alpha(\boldsymbol{r}), \psi_\beta(\boldsymbol{r}')\} = 0, \tag{2.58}$$

$$\{A, B\} = AB + BA. \tag{2.59}$$

To diagonalize the Hamiltonian, we apply the Fourier transformation of operator

$$\psi_\alpha(\boldsymbol{r}) = \frac{1}{\sqrt{V_{\mathrm{vol}}}} \sum_{\boldsymbol{k}} c_{\boldsymbol{k},\alpha} e^{i\boldsymbol{k}\cdot\boldsymbol{r}}, \tag{2.60}$$

where V_{vol} is the volume of a metal. The basis functions satisfy the normalization condition and completeness,

$$\frac{1}{V_{\mathrm{vol}}} \int d\boldsymbol{r}\, e^{i(\boldsymbol{k}-\boldsymbol{k}')\cdot\boldsymbol{r}} = \delta_{\boldsymbol{k},\boldsymbol{k}'}, \tag{2.61}$$

$$\frac{1}{V_{\mathrm{vol}}} \sum_{\boldsymbol{k}} e^{i\boldsymbol{k}\cdot(\boldsymbol{r}-\boldsymbol{r}')} = \frac{1}{(2\pi)^d} \int d\boldsymbol{k}\, e^{i\boldsymbol{k}\cdot(\boldsymbol{r}-\boldsymbol{r}')} = \delta(\boldsymbol{r}-\boldsymbol{r}'), \tag{2.62}$$

where d represents the spatial dimension of the metal. The anticommutation relation of $c_{\boldsymbol{k},\alpha}$ can be derived as

$$\{c_{\boldsymbol{k},\alpha}, c_{\boldsymbol{p},\beta}^\dagger\} = \frac{1}{V_{\mathrm{vol}}} \int d\boldsymbol{r} \int d\boldsymbol{r}'\, \{\psi_\alpha(\boldsymbol{r}), \psi_\beta^\dagger(\boldsymbol{r}')\} e^{-i\boldsymbol{k}\cdot\boldsymbol{r}} e^{i\boldsymbol{p}\cdot\boldsymbol{r}'}, \tag{2.63}$$

$$= \delta_{\alpha,\beta} \frac{1}{V_{\mathrm{vol}}} \int d\boldsymbol{r}\, e^{-i(\boldsymbol{k}-\boldsymbol{p})\cdot\boldsymbol{r}} = \delta_{\alpha,\beta}\delta_{\boldsymbol{k},\boldsymbol{p}}, \tag{2.64}$$

$$\{c_{\boldsymbol{k},\alpha}, c_{\boldsymbol{p},\beta}\} = 0. \tag{2.65}$$

Substituting the Fourier representation, the Hamiltonian becomes

$$\mathcal{H}_0 = \sum_{\alpha=\uparrow,\downarrow} \int d\boldsymbol{r}\, \frac{1}{V_{\mathrm{vol}}} \sum_{\boldsymbol{k},\boldsymbol{k}'} e^{-i\boldsymbol{k}'\cdot\boldsymbol{r}} c_{\boldsymbol{k}',\alpha}^\dagger \left[-\frac{\hbar^2\nabla^2}{2m} - \epsilon_F \right] e^{i\boldsymbol{k}\cdot\boldsymbol{r}} c_{\boldsymbol{k},\alpha}, \tag{2.66}$$

$$= \sum_{\alpha=\uparrow,\downarrow} \frac{1}{V_{\mathrm{vol}}} \int d\boldsymbol{r} \sum_{\boldsymbol{k},\boldsymbol{k}'} e^{i(\boldsymbol{k}-\boldsymbol{k}')\cdot\boldsymbol{r}} c_{\boldsymbol{k}',\alpha}^\dagger\, \xi_{\boldsymbol{k}}\, c_{\boldsymbol{k},\alpha} = \sum_{\alpha=\uparrow,\downarrow} \sum_{\boldsymbol{k}} \xi_{\boldsymbol{k}}\, c_{\boldsymbol{k},\alpha}^\dagger c_{\boldsymbol{k},\alpha}, \tag{2.67}$$

where $\xi_{\boldsymbol{k}} = \hbar^2\boldsymbol{k}^2/(2m) - \epsilon_F$ is the kinetic energy of an electron measured from the Fermi energy $\epsilon_F = \hbar^2 k_F^2/(2m)$.

Let us check the average of number operator given by

$$\langle c^{\dagger}_{\boldsymbol{k},\alpha} c_{\boldsymbol{p},\lambda}\rangle = \frac{1}{\Xi}\mathrm{Tr}[e^{-\beta\mathcal{H}} c^{\dagger}_{\boldsymbol{k},\alpha} c_{\boldsymbol{p},\lambda}] = \frac{1}{\Xi}\mathrm{Tr}[c_{\boldsymbol{p},\lambda} e^{-\beta\mathcal{H}} c^{\dagger}_{\boldsymbol{k},\alpha}], \tag{2.68}$$

$$= \frac{1}{\Xi}\mathrm{Tr}[e^{-\beta\mathcal{H}} I(\beta) c^{\dagger}_{\boldsymbol{k},\alpha}], \tag{2.69}$$

$$I(\beta) = e^{\beta\mathcal{H}} c_{\boldsymbol{p},\lambda} e^{-\beta\mathcal{H}}. \tag{2.70}$$

Since $[\mathcal{H}, c_{\boldsymbol{p},\lambda}] = -\xi_{\boldsymbol{p}} c_{\boldsymbol{p},\lambda}$, it is possible to replace $I(\beta)$ by $c_{\boldsymbol{p},\lambda} e^{-\beta\xi_{\boldsymbol{p}}}$ as discussed in Eq. (2.51). The average is then described as

$$\langle c^{\dagger}_{\boldsymbol{k},\alpha} c_{\boldsymbol{p},\lambda}\rangle = \frac{1}{\Xi}\mathrm{Tr}[e^{-\beta\mathcal{H}} c_{\boldsymbol{p},\lambda} c^{\dagger}_{\boldsymbol{k},\alpha}] e^{-\beta\xi_{\boldsymbol{p}}}, \tag{2.71}$$

$$= \frac{1}{\Xi}\mathrm{Tr}[e^{-\beta\mathcal{H}} (\delta_{\alpha,\lambda}\delta_{\boldsymbol{k},\boldsymbol{p}} - c^{\dagger}_{\boldsymbol{k},\alpha} c_{\boldsymbol{p},\lambda})] e^{-\beta\xi_{\boldsymbol{p}}}, \tag{2.72}$$

$$= \delta_{\alpha,\lambda}\delta_{\boldsymbol{k},\boldsymbol{p}} e^{-\beta\xi_{\boldsymbol{p}}} - \langle c^{\dagger}_{\boldsymbol{k},\alpha} c_{\boldsymbol{p},\lambda}\rangle e^{-\beta\xi_{\boldsymbol{p}}}. \tag{2.73}$$

Therefore, we reach the expression

$$\langle c^{\dagger}_{\boldsymbol{k},\alpha} c_{\boldsymbol{p},\lambda}\rangle = \delta_{\alpha,\lambda}\,\delta_{\boldsymbol{k},\boldsymbol{p}}\, n_F(\xi_{\boldsymbol{k}}), \tag{2.74}$$

$$n_F(\xi_{\boldsymbol{k}}) = \frac{1}{e^{\beta\xi_{\boldsymbol{k}}} + 1} = \frac{1}{2}\left[1 - \tanh\left(\frac{\xi_{\boldsymbol{k}}}{2k_B T}\right)\right], \tag{2.75}$$

where n_F is the Fermi-Dirac distribution function. Equations (2.64) and (2.65) are the anticommutation relations of fermion operators. It is easy to show $\langle c_{\boldsymbol{k},\alpha} c_{\boldsymbol{p},\lambda}\rangle = 0$ and $\langle c_{\boldsymbol{k},\alpha} c^{\dagger}_{\boldsymbol{p},\lambda}\rangle = [1 - n_F(\xi_{\boldsymbol{k}})]\delta_{\alpha,\lambda}\,\delta_{\boldsymbol{k},\boldsymbol{p}}$.

In usual metals such as Au, Al, and Pb, a crystal with volume of 1 cm^3 contains $N_A \sim 10^{23}$ metallic ions. Therefore, the density of free electrons is about 10^{23} cm^{-3}. At $T = 0$, N_A electrons occupy all the possible states from the band bottom up to the Fermi energy ϵ_F. The Fermi energy is typically 10^4 K and is much larger than the room temperature of 300 K. The Fermi velocity $v_F = \hbar k_F/m$ is about 10^8 cm/s which is 1/100 of the speed of light $c = 3\times 10^{10}$ cm/s. Electrons on the Fermi surface have such high kinetic energy due to their statistics. In addition, the degree of the degeneracy at the Fermi energy is large. The quantum mechanics suggests that the large degree of degeneracy in quantum states is a consequence of high symmetry of Hamiltonian. In metals, electrons in conduction bands always interact with another bosonic excitations such as phonon and magnon. In addition, the Coulomb repulsive interaction mediated by photons works between two electrons. Electrons often break the high symmetry spontaneously by using such interactions. As a result, electrons can decrease their ground state energy because the symmetry breaking lifts the degeneracy. Indeed a number of simple metals becomes ferromagnetic or superconducting at a low enough temperature. Figure 2.1 shows the periodic table of elements, where elements colored in red or pink are superconductors and those in blue are ferromagnets. The table suggests that superconductivity at low temper-

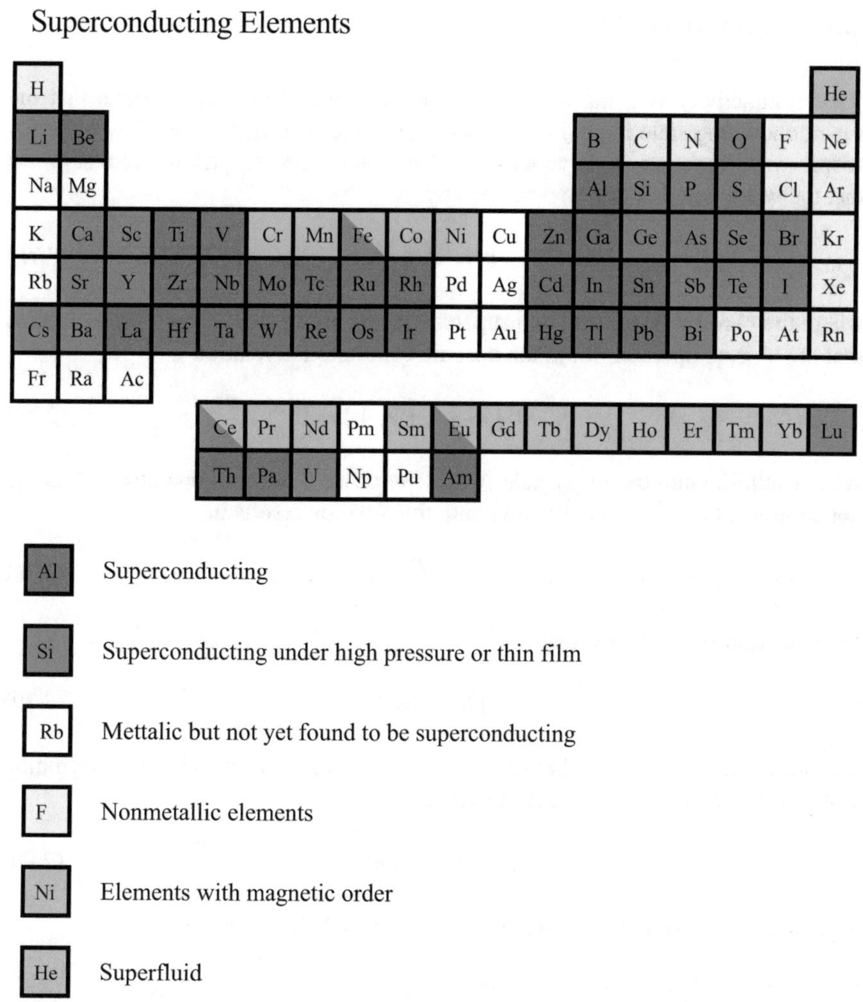

Fig. 2.1 The periodic table of elements. Elements colored in red or pink are superconductors (Hamlin 2015)

atures is a common feature of many metals. The short-range repulsive interaction between the electrons breaks time-reversal symmetry, resulting in the appearance of ferromagnetism. The coupling between electrons and the lattice distortion breaks translational symmetry, resulting in the appearance of charge density waves. The phase transition of the second kind describes well the appearance of such ordered phases at a low temperature. The transition to superconducting state is also a phase transition of the second kind which is caused by the attractive interaction between two electrons. The superconducting state breaks gauge symmetry spontaneously. In what follows, we will see the physical consequence of the breaking gauge symmetry.

2.3 Coherent State

Superconductivity is a macroscopic quantum phenomenon. This means that the superconducting state including N_{A} electrons could be described well by a macroscopic wave function Ψ. Feynman's explanation of the Josephson effect suggests that the wave function can be described by

$$\Psi = \sqrt{N}e^{i\theta}, \tag{2.76}$$

where the phase θ is uniform in a superconductor. In the field theory, let us assume that the Ψ is an operator obeys the bosonic commutation relation

$$[\Psi, \Psi^\dagger] = 1, \quad [\Psi, \Psi] = 0. \tag{2.77}$$

As a result, the number of particle N and the phase θ are also operators which do not commute to each other. The commutation relation results in

$$[\Psi, \Psi^\dagger] = \sqrt{N}e^{i\theta}e^{-i\theta}\sqrt{N} - e^{-i\theta}\sqrt{N}\sqrt{N}e^{i\theta} = N - e^{-i\theta}Ne^{i\theta}. \tag{2.78}$$

We will show that the relation

$$[N, \theta] = i \tag{2.79}$$

is a sufficient condition for the right-hand side of Eq. (2.78) being unity. Beginning with Eq. (2.79), we first prove the relation

$$[N, \theta^m] = m\, i\, \theta^{m-1}. \tag{2.80}$$

Equation (2.79) corresponds to the case of $m = 1$. At $m \to m + 1$,

$$[N, \theta^{m+1}] = N\theta^m\theta - \theta^{m+1}N = (mi\theta^{m-1} + \theta^m N)\theta - \theta^{m+1}N, \tag{2.81}$$

$$= mi\theta^m + \theta^m(i + \theta N) - \theta^{m+1}N = (m+1)i\theta^m. \tag{2.82}$$

Therefore, a commutator becomes

$$[N, e^{-i\theta}] = \left[N, \sum_{m=0}^{\infty}\frac{(-1)^m}{m!}\theta^m\right] = \sum_{m=0}^{\infty}\frac{(-1)^m}{m!}\, m\, i\, \theta^{m-1} = i\frac{d}{d\theta}\sum_{m=0}^{\infty}\frac{(-1)^m}{m!}\theta^m, \tag{2.83}$$

$$= i\frac{d}{d\theta}e^{-i\theta} = e^{-i\theta}. \tag{2.84}$$

Thus, Eq. (2.78) results in $[\Psi, \Psi^\dagger] = 1$. Although Eq. (2.79) is only a sufficient condition, it has an important physical meaning. Equation (2.79) suggests that N

and θ are canonical conjugate to each other. Thus, it is impossible to fix the two values simultaneously. Namely if we fix the number of electrons, the phase is unfixed. On the other hand, when we fix the phase, the number of electrons is unfixed. The former is realized in the normal metallic state, whereas the latter corresponds to the superconducting state.

To imagine the phase fixed state, let us consider the classical coherent light so-called laser beam. The state of light is represented using the bosonic annihilation (creation) operator of a photon a ($a^\dagger$) with the vacuum $a|0\rangle = 0$. The coherent state $|\mathrm{C}\rangle$ is defined as an eigenstate of the annihilation operator,

$$a|\mathrm{C}\rangle = \alpha|\mathrm{C}\rangle, \tag{2.85}$$

where the eigenvalue $\alpha = |\alpha|e^{i\varphi_\alpha}$ is a complex number. The coherent state is decomposed into a series of the number states of photons,

$$|\mathrm{C}\rangle = \sum_{n=0}^{\infty} c_n|n\rangle, \quad |n\rangle = \frac{(a^\dagger)^n}{\sqrt{n!}}|0\rangle. \tag{2.86}$$

According to the definition, we find

$$a|\mathrm{C}\rangle = \sum_{n=1}^{\infty} c_n\sqrt{n}|n-1\rangle = \alpha\sum_{m=0}^{\infty} c_m|m\rangle = \alpha\sum_{n=1}^{\infty} c_{n-1}|n-1\rangle, \tag{2.87}$$

and obtain the relation $c_n\sqrt{n} = c_{n-1}\alpha$. By applying the recursive relation repeatedly, the coefficients of the series is represented as

$$c_n = \frac{\alpha}{\sqrt{n}}c_{n-1} = \frac{\alpha}{\sqrt{n}}\frac{\alpha}{\sqrt{n-1}}c_{n-2} = \cdots = \frac{\alpha^n}{\sqrt{n!}}c_0, \tag{2.88}$$

$$|\mathrm{C}\rangle = c_0\sum_{n=0}^{\infty}\frac{\alpha^n}{\sqrt{n!}}|n\rangle. \tag{2.89}$$

A constant c_0 is determined by

$$\langle\mathrm{C}|\mathrm{C}\rangle = |c_0|^2\sum_{n=0}^{\infty}\sum_{m=0}^{\infty}\frac{(\alpha^*)^n}{\sqrt{n!}}\frac{\alpha^m}{\sqrt{m!}}\langle n|m\rangle = |c_0|^2\sum_{n=0}^{\infty}\frac{|\alpha|^{2n}}{n!} = |c_0|^2 e^{|\alpha|^2} = 1. \tag{2.90}$$

The coherent state is then described as

$$\begin{aligned} |\mathrm{C}\rangle =& e^{-\frac{1}{2}|\alpha|^2}\sum_{n=0}^{\infty}\frac{\alpha^n}{\sqrt{n!}}|n\rangle = e^{-\frac{1}{2}|\alpha|^2}\sum_{n=0}^{\infty}\frac{(\alpha a)^n}{n!}|0\rangle, \\ =& e^{-\frac{1}{2}|\alpha|^2}\left[1 + (|\alpha|\,e^{i\varphi_\alpha}a^\dagger) + \frac{1}{2}(|\alpha|\,e^{i\varphi_\alpha}a^\dagger)^2 + \cdots + \frac{1}{n!}(|\alpha|\,e^{i\varphi_\alpha}a^\dagger)^n + \cdots\right]|0\rangle. \end{aligned} \tag{2.91}$$

The number of photons is not fixed in the coherent state. In addition, all photons are generated on the vacuum with the same phase of φ_α. Such a coherent state is realized in the classical coherent light.

Ginzburg-Landau theory of superconductivity describes phenomenologically the equation of motion for the wave function $\Psi(\boldsymbol{r}) = \sqrt{n(\boldsymbol{r})}\, e^{i\theta(\boldsymbol{r})}$. In the microscopic theory, the normal ground state of free electrons in a metal is described by

$$|\mathrm{N}\rangle = \prod_{|\boldsymbol{k}|\leq k_F} c^\dagger_{\boldsymbol{k},\uparrow} c^\dagger_{\boldsymbol{k},\downarrow} |0\rangle. \tag{2.92}$$

When the number of states for $|\boldsymbol{k}| \leq k_F$ is N, the number of electron is fixed at $2N$. According to the BCS theory (Bardeen et al. 1957), the superconducting state is described by

$$|\mathrm{S}\rangle = \prod_{|\boldsymbol{k}|\leq k_F} \left(u_{\boldsymbol{k}} + v_{\boldsymbol{k}} e^{i\varphi} c^\dagger_{-\boldsymbol{k},\downarrow} c^\dagger_{\boldsymbol{k},\uparrow} \right) |0\rangle, \tag{2.93}$$

with $u_{\boldsymbol{k}}^2 + v_{\boldsymbol{k}}^2 = 1$. A pair of two electrons are created with a phase of φ independent of $\boldsymbol{k}$. The superconducting condensate is described by the coherent superposition of such electron pairs. It is clear that the number of electrons is not fixed in the superconducting state in Eq. (2.93).

References

Bardeen, J., Cooper, L.N., Schrieffer, J.R.: Phys. Rev. **108**, 1175 (1957). https://doi.org/10.1103/PhysRev.108.1175

Hamlin, J.: Phys. C: Supercond. Appl. **514**, 59 (2015). https://doi.org/10.1016/j.physc.2015.02.032, http://www.sciencedirect.com/science/article/pii/S0921453415000593. Superconducting Materials: Conventional, Unconventional and Undetermined

Chapter 3
Mean-Field Theory of Superconductivity

Abstract We derive the mean-field Hamiltonian for superconductivity with paying careful attention to the pair potential and its physical consequence. The macroscopic wave function is introduced to describe the superconducting condensate. The magnetic properties of a superconductor are described by solving the Maxwell equation with a current expression in terms of the macroscopic wave function.

3.1 Pairing Hamiltonian

The mechanism of superconductivity in metals was explained by Bardeen, Cooper, and Schrieffer (BCS) in 1957. Although the variational wave function method was used in the original paper, we often use the mean-field theory to describe the superconducting condensate today. The mean-field theory is advantageous when we calculate various physical values and generalize the theory to inhomogeneous superconductors or superconducting junctions. The energy of the superconducting state in Eq. (2.93) is lower than that of the normal state in Eq. (2.92) when a weak attractive interaction works between two electrons on the Fermi surface (Bardeen et al. 1957). BCS showed that the Fermi surface is fragile in the presence of the attractive interactions between two electrons. In a metal, the repulsive Coulomb interaction mediated by a photon always works between two electrons. In the Coulomb gauge, the coulomb interaction is instantaneous. The attractive interaction, on the other hand, is not instantaneous because the velocity of phonon $v_{ph} \approx 100$ m/s is much slower than v_F. Thus such retardation effect in the attractive interaction enables two electrons to form a pair. The classical image of the pairing mechanism may be explained as follows. An electron comes to a certain place (say $\boldsymbol{r}_0$) in a metal at a certain time (say $t = 0$) and attracts ions surrounding it. The coming electron stays around $\boldsymbol{r}_0$ within a time scale given by $t_e \sim \hbar/\epsilon_F$. On the other hand, it takes $t_{ph} \sim 1/\omega_D \gg t_e$ for ions to move toward $\boldsymbol{r}_0$, where ω_D is called Debye frequency. At $t = t_{ph}$, the electron has already moved away from $\boldsymbol{r}_0$. But remaining ions charge the place positively, which attracts another electron. Two electrons on the Fermi surface form a Cooper pair due to such an attractive interaction nonlocal in time. The classical image of the interaction is illustrated in Fig. 3.1.

Y. Asano, *Andreev Reflection in Superconducting Junctions*,
SpringerBriefs in Physics, https://doi.org/10.1007/978-981-16-4165-7_3

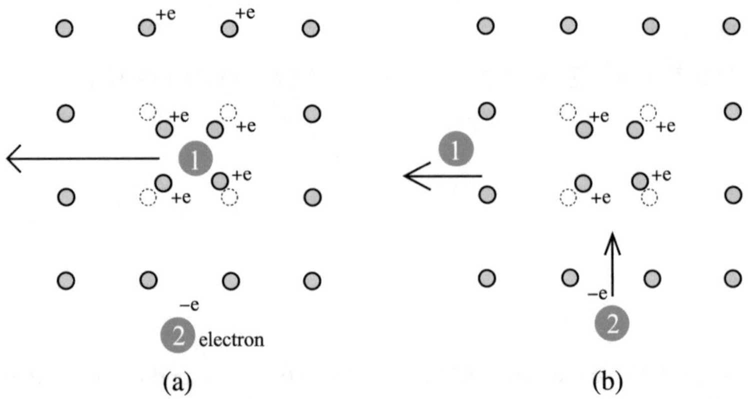

Fig. 3.1 The schematic image for electron-phonon interaction. **a** The first electron modifies the lattice structure locally and leaves away. **b** The local positive charge attracts the second electron

The interaction between two electrons is described by the Hamiltonian

$$\mathcal{H}_I = \frac{1}{2}\sum_{\alpha,\beta}\int d\boldsymbol{r}\int d\boldsymbol{r}'\psi^\dagger_\alpha(\boldsymbol{r})\,\psi^\dagger_\beta(\boldsymbol{r}')\,V(\boldsymbol{r}-\boldsymbol{r}')\,\psi_\beta(\boldsymbol{r}')\,\psi_\alpha(\boldsymbol{r}). \tag{3.1}$$

The total Hamiltonian we consider is $\mathcal{H} = \mathcal{H}_0 + \mathcal{H}_I$, where $\mathcal{H}_0$ is given in Eq. (2.57). By applying the Fourier transformation in Eq. (2.60), we obtain $\mathcal{H}_0$ in Eq. (2.67) and

$$\mathcal{H}_I = \frac{1}{2V_{\mathrm{vol}}}\sum_{\alpha,\beta}\sum_{\boldsymbol{k}_1,\boldsymbol{k}_2,\boldsymbol{q}} V_{\boldsymbol{q}}\, c^\dagger_{\boldsymbol{k}_1+\boldsymbol{q},\alpha}\, c^\dagger_{\boldsymbol{k}_2-\boldsymbol{q},\beta}\, c_{\boldsymbol{k}_2,\beta}\, c_{\boldsymbol{k}_1,\alpha}, \tag{3.2}$$

$$V(\boldsymbol{r}) = \frac{1}{V_{\mathrm{vol}}}\sum_{\boldsymbol{q}} V_{\boldsymbol{q}} e^{i\boldsymbol{q}\cdot\boldsymbol{r}}. \tag{3.3}$$

We assume that V_q satisfies following conditions,

1. The interaction is attractive and independent of $\boldsymbol{q}$, (i.e., $V_{\boldsymbol{q}} = -g$).
2. The interaction works between two electrons at $\boldsymbol{k}_2 = -\boldsymbol{k}_1$ and $\beta = \bar{\alpha}$.
3. The scattering event by the interaction happens in a small energy window near the Fermi level characterized by $\hbar\omega_D$.

Under these conditions, the interaction Hamiltonian becomes

$$\mathcal{H}_I = \frac{1}{V_{\mathrm{vol}}}\sum_{\boldsymbol{k},\boldsymbol{k}'} V(\boldsymbol{k},\boldsymbol{k}')\, c^\dagger_{-\boldsymbol{k}',\downarrow}\, c^\dagger_{\boldsymbol{k}',\uparrow}\, c_{\boldsymbol{k},\uparrow}\, c_{-\boldsymbol{k},\downarrow}, \tag{3.4}$$

$$V(\boldsymbol{k},\boldsymbol{k}') = -\,g\;\Theta(-|\xi_{\boldsymbol{k}}| + \hbar\omega_D)\Theta(-|\xi_{\boldsymbol{k}'}| + \hbar\omega_D), \tag{3.5}$$

where $g > 0$ is a constant. The total Hamiltonian

$$\mathcal{H} = \sum_{k,\alpha} \xi_k \, c^\dagger_{k,\alpha} \, c_{k,\alpha} - \frac{g}{V_{\text{vol}}} \sum_{k}{}' \sum_{k'}{}' c^\dagger_{-k',\downarrow} \, c^\dagger_{k',\uparrow} \, c_{k,\uparrow} \, c_{-k,\downarrow}, \tag{3.6}$$

$$\sum_k{}' = \sum_k \Theta(-|\xi_k| + \hbar\omega_D) \tag{3.7}$$

is called pairing Hamiltonian. The interaction at the final expression in real space is given by $V(\boldsymbol{r} - \boldsymbol{r}') = -g\delta(\boldsymbol{r} - \boldsymbol{r}')$ which is short range in space and instantaneous in time. These properties are different qualitatively from the effective interaction mediated by a phonon. In this sense, Eq. (3.6) is a model Hamiltonian that describes many-electron states in the presence of a specialized attractive interaction between two electrons on the Fermi surface. As we will see below, however, the mean-field theory for Eq. (3.6) explains well the superconducting phenomena observed in experiments. This suggests that the attractive interaction between two electrons is indispensable to superconductivity and its dependence on space-time plays only a minor role.

3.2 Mean-Field Approximation

The interaction term in Eq. (3.6) is nonlinear. It is impossible to have an exact solution of the ground state. Here, instead of solving Eq. (3.6) exactly, we apply the mean-field approximation to the interaction term. We introduce the average of operators

$$\Delta \, e^{i\varphi} \equiv \frac{g}{V_{\text{vol}}} \sum_k{}' \langle c_{k,\uparrow} \, c_{-k,\downarrow} \rangle, \tag{3.8}$$

which is called pair potential and plays a central role in describing superconductivity. Generally speaking, the right-hand side is a complex number. Thus, the left-hand side is decomposed into the amplitude and the phase of the pair potential. Before solving the mean-field Hamiltonian, the meaning of the pair potential should be clarified. The product of two annihilation operators decreases the number of electrons by two. If we consider a ground state in which the number of electrons N is fixed, the quantum average results in

$$\langle N | c_{k,\uparrow} \, c_{-k,\downarrow} | N \rangle \propto \langle N | N - 2 \rangle = 0. \tag{3.9}$$

If the pair potential in Eq. (3.8) takes on a nonzero value, the number of electrons is not fixed in such a ground state. The coherent state of the laser light can be a possible candidate of the ground state. Let us replace a product of electron operators by a boson operator phenomenologically as $a_k = c_{k,\uparrow} \, c_{-k,\downarrow}$. Since $a_k |\text{C}\rangle \propto |\text{C}\rangle$, the pair potential can have a nonzero value in the coherent state. Thus, Eq. (3.8) automatically determines the character of the ground state.

Our next task is to linearize the BCS Hamiltonian by using the mean field in Eq. (3.8),

$$c_{k,\uparrow}\, c_{-k,\downarrow} = \langle c_{k,\uparrow}\, c_{-k,\downarrow}\rangle + \left[c_{k,\uparrow}\, c_{-k,\downarrow} - \langle c_{k,\uparrow}\, c_{-k,\downarrow}\rangle\right], \tag{3.10}$$

$$c^\dagger_{-k,\downarrow}\, c^\dagger_{k,\uparrow} = \langle c^\dagger_{-k,\downarrow}\, c^\dagger_{k,\uparrow}\rangle + \left[c^\dagger_{-k,\downarrow}\, c^\dagger_{k,\uparrow} - \langle c^\dagger_{-k,\downarrow}\, c^\dagger_{k,\uparrow}\rangle\right]. \tag{3.11}$$

These are the identities. The first term on the right-hand-side is the average on the left-hand side. The second term is considered to be the fluctuations from the average. In the mean-field theory, we assume that the fluctuations are much smaller than the average and expand the nonlinear term within the first order of the fluctuations. The interaction term becomes

$$\sum_{k'}{}' c^\dagger_{-k',\downarrow}\, c^\dagger_{k',\uparrow} \sum_{k}{}' c_{k,\uparrow}\, c_{-k,\downarrow}, \tag{3.12}$$

$$= \sum_{k'}{}' \langle c^\dagger_{-k',\downarrow}\, c^\dagger_{k',\uparrow}\rangle + \left[c^\dagger_{-k',\downarrow}\, c^\dagger_{k',\uparrow} - \langle c^\dagger_{-k',\downarrow}\, c^\dagger_{k',\uparrow}\rangle\right]$$
$$\times \sum_{k}{}' \langle c_{k,\uparrow}\, c_{-k,\downarrow}\rangle + \left[c_{k,\uparrow}\, c_{-k,\downarrow} - \langle c_{k,\uparrow}\, c_{-k,\downarrow}\rangle\right], \tag{3.13}$$

$$= \left\{\frac{V_{\mathrm{vol}}\Delta e^{-i\varphi}}{g} + \left[\sum_{k'}{}' c^\dagger_{-k',\downarrow}\, c^\dagger_{k',\uparrow} - \frac{V_{\mathrm{vol}}\Delta e^{-i\varphi}}{g}\right]\right\}$$
$$\times \left\{\frac{V_{\mathrm{vol}}\Delta e^{i\varphi}}{g} + \left[\sum_{k}{}' c_{k,\uparrow}\, c_{-k,\downarrow} - \frac{V_{\mathrm{vol}}\Delta e^{i\varphi}}{g}\right]\right\}, \tag{3.14}$$

$$\approx -\frac{V_{\mathrm{vol}}^2\Delta^2}{g^2} + \frac{V_{\mathrm{vol}}}{g}\sum_{k}{}' \left[\Delta e^{-i\varphi} c_{k,\uparrow}\, c_{-k,\downarrow} + \Delta e^{i\varphi} c^\dagger_{-k,\downarrow}\, c^\dagger_{k,\uparrow}\right]. \tag{3.15}$$

On the way to the derivation, we have used the complex conjugation of the pair potential

$$\Delta\, e^{-i\varphi} = \frac{g}{V_{\mathrm{vol}}}\sum_{k}{}' \langle c^\dagger_{-k,\downarrow}\, c^\dagger_{k,\uparrow}\rangle. \tag{3.16}$$

The mean-field Hamiltonian for superconductivity has the form

$$\mathcal{H}_{\mathrm{MF}} = \sum_{k,\alpha} \xi_k\, c^\dagger_{k,\alpha}\, c_{k,\alpha} - \sum_{k}{}' \left[\Delta e^{-i\varphi} c_{k,\uparrow}\, c_{-k,\downarrow} + \Delta e^{i\varphi} c^\dagger_{-k,\downarrow}\, c^\dagger_{k,\uparrow}\right] + \frac{V_{\mathrm{vol}}\Delta^2}{g}, \tag{3.17}$$

$$= \sum_{k} \left[c^\dagger_{k,\uparrow}, c_{-k,\downarrow}\right] \begin{bmatrix} \xi_k & \Delta\, e^{i\varphi} \\ \Delta\, e^{-i\varphi} & -\xi_k \end{bmatrix} \begin{bmatrix} c_{k,\uparrow} \\ c^\dagger_{-k,\downarrow} \end{bmatrix} + \frac{V_{\mathrm{vol}}\Delta^2}{g}. \tag{3.18}$$

The high-energy cut off in the summation should be consider if necessary. To diagonalize the mean-field Hamiltonian, we solve the eigenequation,

$$\begin{bmatrix} \xi_k & \Delta e^{i\varphi} \\ \Delta e^{-i\varphi} & -\xi_k \end{bmatrix} \begin{bmatrix} a \\ b \end{bmatrix} = E \begin{bmatrix} a \\ b \end{bmatrix}. \tag{3.19}$$

The solutions are given by

$$\begin{bmatrix} u_k \\ v_k e^{-i\varphi} \end{bmatrix}, \quad \begin{bmatrix} -v_k e^{i\varphi} \\ u_k \end{bmatrix}, \tag{3.20}$$

for $E = E_k$ and $E = -E_k$, respectively. Here, we define

$$E_k = \sqrt{\xi_k^2 + \Delta^2}, \quad u_k = \sqrt{\frac{1}{2}\left(1 + \frac{\xi_k}{E_k}\right)}, \quad v_k = \sqrt{\frac{1}{2}\left(1 - \frac{\xi_k}{E_k}\right)}. \tag{3.21}$$

The results are summarized as

$$\begin{bmatrix} \xi_k & \Delta\, e^{i\varphi} \\ \Delta\, e^{-i\varphi} & -\xi^*_{-k} \end{bmatrix} = \begin{bmatrix} u_k & -v_k e^{i\varphi} \\ v_k e^{-i\varphi} & u_k \end{bmatrix} \begin{bmatrix} E_k & 0 \\ 0 & -E_k \end{bmatrix} \begin{bmatrix} u_k & v_k e^{i\varphi} \\ -v_k e^{-i\varphi} & u_k \end{bmatrix}. \tag{3.22}$$

By substituting the results into Eq. (3.18), the Hamiltonian is diagonalized as

$$\mathcal{H}_{\text{MF}} = \sum_k \left[\gamma^\dagger_{k,\uparrow}, \gamma_{-k,\downarrow}\right] \begin{bmatrix} E_k & 0 \\ 0 & -E_{-k} \end{bmatrix} \begin{bmatrix} \gamma_{k,\uparrow} \\ \gamma^\dagger_{-k,\downarrow} \end{bmatrix} + \frac{V_{\text{vol}}\Delta^2}{g}, \tag{3.23}$$

$$= \sum_k E_k(\gamma^\dagger_{k,\uparrow}\gamma_{k,\uparrow} + \gamma^\dagger_{-k,\downarrow}\gamma_{-k,\downarrow} - 1) + \frac{V_{\text{vol}}\Delta^2}{g}, \tag{3.24}$$

with

$$\begin{bmatrix} \gamma_{k,\uparrow} \\ \gamma^\dagger_{-k,\downarrow} \end{bmatrix} = \begin{bmatrix} u_k & v_k e^{i\varphi} \\ -v_k e^{-i\varphi} & u_k \end{bmatrix} \begin{bmatrix} c_{k,\uparrow} \\ c^\dagger_{-k,\downarrow} \end{bmatrix}, \tag{3.25}$$

$$\begin{bmatrix} c_{k,\uparrow} \\ c^\dagger_{-k,\downarrow} \end{bmatrix} = \begin{bmatrix} u_k & -v_k e^{i\varphi} \\ v_k e^{-i\varphi} & u_k \end{bmatrix} \begin{bmatrix} \gamma_{k,\uparrow} \\ \gamma^\dagger_{-k,\downarrow} \end{bmatrix}. \tag{3.26}$$

The last relationships are called Bogoliubov transformation (Bogoliubov 1958a, b). The operator $\gamma_{k,\alpha}$ is the annihilation operator of a Bogoliubov quasiparticle and obeys the fermionic anticommutation relations

$$\{\gamma_{k,\alpha}, \gamma^\dagger_{p,\beta}\} = \delta_{k,p}\delta_{\alpha,\beta}, \quad \{\gamma_{k,\alpha}, \gamma_{p,\beta}\} = 0, \quad \gamma_{k,\alpha}|\tilde{0}\rangle = 0. \tag{3.27}$$

The last equation defines the "vacuum" of the Bogoliubov quasiparticle, where $|\tilde{0}\rangle$ represents the superconducting ground state. The creation of a Bogoliubov quasiparticle describes the elementary excitation from the superconducting ground state.

3.3 Gap Equation and Thermodynamic Properties

The remaining task is to determine the pair potential Δ in a self-consistent way. By the definition of the pair potential Eq. (3.8) and the Bogoliubov transformation in Eq. (3.26), we obtain

$$\Delta\, e^{i\varphi} = \frac{g}{V_{\text{vol}}} \sum_{\boldsymbol{k}}{}' \left\langle (u_{\boldsymbol{k}} \gamma_{\boldsymbol{k},\uparrow} - v_{\boldsymbol{k}} e^{i\varphi} \gamma^{\dagger}_{-\boldsymbol{k},\downarrow})(v_{\boldsymbol{k}} e^{i\varphi} \gamma^{\dagger}_{\boldsymbol{k},\uparrow} + u_{\boldsymbol{k}} \gamma_{-\boldsymbol{k},\downarrow}) \right\rangle, \tag{3.28}$$

$$= \frac{g}{V_{\text{vol}}} \sum_{\boldsymbol{k}}{}' u_{\boldsymbol{k}} v_{\boldsymbol{k}} e^{i\varphi} \left(1 - 2n_F(E_{\boldsymbol{k}})\right) = \frac{g}{V_{\text{vol}}} \sum_{\boldsymbol{k}}{}' \frac{\Delta\, e^{i\varphi}}{2E_{\boldsymbol{k}}} \tanh\left(\frac{E_{\boldsymbol{k}}}{2k_B T}\right). \tag{3.29}$$

On the way to the last line, we have used the relations,

$$\langle \gamma^{\dagger}_{\boldsymbol{k},\alpha} \gamma_{\boldsymbol{p},\beta} \rangle = n_F(E_{\boldsymbol{k}})\, \delta_{\boldsymbol{k},\boldsymbol{p}}\, \delta_{\alpha,\beta}, \quad \langle \gamma_{\boldsymbol{k},\alpha} \gamma_{\boldsymbol{p},\beta} \rangle = 0. \tag{3.30}$$

The amplitude of Δ should be determined by solving Eq. (3.29). Thus Eq. (3.29) is called gap equation. We note that the equation has a solution only for $g > 0$ (attractive interaction). To proceed the calculation, we introduce the density of states per volume per spin,

$$N(\xi) = \frac{1}{V_{\text{vol}}} \sum_{\boldsymbol{k}} \delta(\xi - \xi_{\boldsymbol{k}}). \tag{3.31}$$

The gap equation becomes

$$1 = g \int_0^{\hbar\omega_D} d\xi\, \frac{N(\xi)}{\sqrt{\xi^2 + \Delta^2}} \tanh\left(\frac{\sqrt{\xi^2 + \Delta^2}}{2k_B T}\right). \tag{3.32}$$

At zero temperature $T = 0$, the integral can be carried out as

$$1 = g \int_0^{\hbar\omega_D} d\xi\, \frac{N(\xi)}{\sqrt{\xi^2 + \Delta_0^2}} \simeq g N_0 \int_0^{\hbar\omega_D} \frac{d\xi}{\sqrt{\xi^2 + \Delta_0^2}}, \tag{3.33}$$

$$= g N_0 \ln\left[\frac{\hbar\omega_D}{\Delta_0} + \sqrt{\left(\frac{\hbar\omega_D}{\Delta_0}\right)^2 + 1}\,\right] \simeq g N_0 \ln(2\hbar\omega_D/\Delta_0), \tag{3.34}$$

where N_0 is the density of states at the Fermi level per spin per unit volume. The amplitude of the pair potential at zero temperature Δ_0 is then described as

$$\Delta_0 = 2\hbar\,\omega_D\, e^{-\frac{1}{gN_0}}. \tag{3.35}$$

As discussed in a number of textbooks, it is impossible to expand the right-hand side with respect to the small parameter $gN_0 \ll 1$, which implies the instability of the Fermi surface in the presence of a weak attractive interaction. Just below the transition temperature T_c, it is possible to put $\Delta \to 0$ into the gap equation,

$$\begin{aligned} 1 =& gN_0 \int_0^{\hbar\omega_D} \frac{d\xi}{\xi} \tanh\left(\frac{\xi}{2k_BT_c}\right), && (3.36)\\ =& gN_0 \ln\left(\frac{\hbar\omega_D}{2k_BT_c}\right) \tanh\left(\frac{\hbar\omega_D}{2k_BT_c}\right) - gN_0 \int_0^{\infty} d\xi\, \ln(\xi)\, \cosh^{-2}(\xi), && (3.37)\\ \simeq& gN_0\left[\ln\left(\frac{\hbar\omega_D}{2k_BT_c}\right) + \ln\left(\frac{4\gamma_0}{\pi}\right)\right] = gN_0 \ln\left(\frac{2\hbar\omega_D\gamma_0}{\pi k_BT_c}\right). && (3.38) \end{aligned}$$

On the way to the last line, we considered $\hbar\omega_D \gg T_c$ and used

$$\begin{aligned} \int_0^{\infty} dx\, \ln(x) \cosh^{-2}(x) =& \log\left(\frac{4\gamma_0}{\pi}\right), && (3.39)\\ \ln\gamma_0 \simeq& 0.577 \qquad \text{Euler constant.} && (3.40) \end{aligned}$$

From the last line, we obtain

$$T_c = \frac{2\hbar\omega_D\gamma_0}{\pi k_B} e^{-1/g N_0} = \frac{\gamma_0}{\pi k_B}\Delta_0. \tag{3.41}$$

The relation of $2\Delta_0 = 3.5k_BT_c$ has been confirmed in a number of metallic superconductors. Figure 3.2 shows the dependence of the pair potential on temperature. As shown in Appendix A, the pair potential just below T_c depends on temperature as

$$\Delta = \pi k_B T_c \sqrt{\frac{8}{7\zeta(3)}}\sqrt{\frac{T_c - T}{T_c}}. \tag{3.42}$$

By using the energy of superconducting state in Eq. (3.24), it is possible to calculate the partition function and the specific heat of a superconductor,

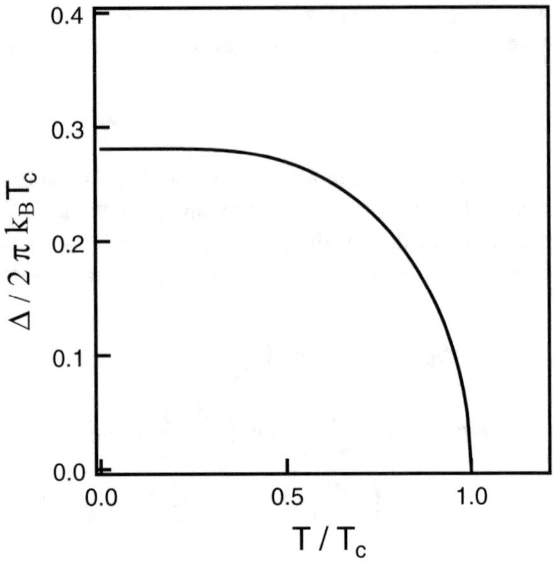

Fig. 3.2 The amplitude of the pair potential is plotted as a function of temperature. Here, we solved the gap equation numerically

$$\Xi_S = \sum_{n_{k,\sigma}=0,1} \exp\left[-\left\{\sum_{k,\sigma} E_k(n_{k,\sigma}-1) + V_{\rm vol}\Delta^2/g\right\}/k_BT\right], \tag{3.43}$$

$$= \exp(-V_{\rm vol}\Delta^2/gk_BT)\prod_k e^{E_k/k_BT}\prod_k \left(1+e^{-E_k/k_BT}\right)^2. \tag{3.44}$$

The thermodynamic potential of the superconducting state per volume is represented by

$$\Omega_S = -k_BT\frac{1}{V_{\rm vol}}\log\Xi_S = \frac{\Delta^2}{g} - \frac{1}{V_{\rm vol}}\sum_k\left[E_k + k_BT\log\left(1+e^{-E_k/k_BT}\right)^2\right]. \tag{3.45}$$

The specific heat is then calculated as

$$S_S = -\frac{\partial\Omega_S}{\partial T} = -\frac{2k_B}{V_{\rm vol}}\sum_k (1-n_F)\log(1-n_F) + n_F\log n_F, \tag{3.46}$$

$$C_S = T\frac{\partial S_S}{\partial T} = \frac{2k_B}{V_{\rm vol}}\sum_k E_k\frac{\partial n_F}{\partial T} = \frac{2k_B}{V_{\rm vol}}\sum_k\left(-\frac{\partial n_F}{\partial E_k}\right)\left(\frac{E_k^2}{T} - \frac{1}{2}\frac{\partial E_k^2}{\partial T}\right), \tag{3.47}$$

$$n_F = \frac{1}{2}\left[1-\tanh\left(\frac{E_k}{2k_BT}\right)\right]. \tag{3.48}$$

At the transition temperature, the specific heat is discontinuous, which implies the rapid decrease of entropy at $T = T_c$. The difference in the specific heats per volume is described as

$$\Delta C = C_S - C_N = 2k_B \int d\xi N(\xi) \left(-\frac{\partial f(\xi)}{\partial \xi}\right)\left(-\frac{1}{2}\frac{\partial \Delta^2}{\partial T}\right), \tag{3.49}$$

$$= N_0 \frac{8\pi^2}{7\zeta(3)} k_B^2 T_c. \tag{3.50}$$

By using the specific heat in the normal state at $T = T_c$

$$C_N = \frac{2\pi^2}{3} N_0 k_B^2 T_c, \tag{3.51}$$

we obtain

$$\frac{\Delta C}{C_N(T_c)} = \frac{12}{7\zeta(3)} \approx 1.42. \tag{3.52}$$

The left hand-side of the equation is a universal value independent of materials.

The free energy in the normal state Ω_N is obtained by putting $\Delta \to 0$ in Ω_S. The difference between the free energies corresponds to the condensation energy of the superconducting state,

$$\Omega_S - \Omega_N = \frac{1}{V_{\mathrm{vol}}} \sum_{\boldsymbol{k}} \times \left[\frac{\Delta^2}{2E_{\boldsymbol{k}}} \tanh\left(\frac{E_{\boldsymbol{k}}}{2k_B T}\right) + |\xi_{\boldsymbol{k}}| - E_{\boldsymbol{k}} - 2k_B T \log \frac{1 + e^{-E_{\boldsymbol{k}}/k_B T}}{1 + e^{-|\xi_{\boldsymbol{k}}|/k_B T}} \right], \tag{3.53}$$

where we have used the gap equation. At $T = 0$, the condensation energy is calculated to be

$$\Omega_S - \Omega_N = 2N_0 \int_0^\infty d\xi \left[\frac{\Delta_0^2}{2E} + \xi - E \right] = -\frac{1}{2} N_0 \Delta_0^2 = -\frac{H_c^2}{8\pi}. \tag{3.54}$$

The last equation relates the condensation energy to the thermodynamic critical field.

3.4 Cooper Pair

The pair correlation function is a two-point function in real space,

$$F(\boldsymbol{r}_1,\boldsymbol{r}_2) = \langle \psi_\uparrow(\boldsymbol{r}_1)\,\psi_\downarrow(\boldsymbol{r}_2)\rangle = \frac{1}{V_{\text{vol}}}\sum_{\boldsymbol{k},\boldsymbol{k}'}{}' \langle c_{\boldsymbol{k},\uparrow}\, c_{\boldsymbol{k}',\downarrow}\rangle e^{i\boldsymbol{k}\cdot\boldsymbol{r}_1} e^{i\boldsymbol{k}'\cdot\boldsymbol{r}_2}, \tag{3.55}$$

$$= \frac{1}{V_{\text{vol}}}\sum_{\boldsymbol{k},\boldsymbol{k}'}{}' \langle c_{\boldsymbol{k},\uparrow}\, c_{\boldsymbol{k}',\downarrow}\rangle e^{i(\boldsymbol{k}+\boldsymbol{k}')\cdot\boldsymbol{r}} e^{i(\boldsymbol{k}-\boldsymbol{k}')\cdot\boldsymbol{\rho}/2}, \tag{3.56}$$

where we introduce the coordinate

$$\boldsymbol{r} = \frac{\boldsymbol{r}_1+\boldsymbol{r}_2}{2}, \quad \boldsymbol{\rho} = \boldsymbol{r}_1 - \boldsymbol{r}_2. \tag{3.57}$$

In a uniform superconductor, the pair correlation function is independent of $\boldsymbol{r}$, which requires $\boldsymbol{k}' = -\boldsymbol{k}$. As a result, correlation function depends only on the relative coordinate as

$$F(\boldsymbol{\rho}) = \frac{1}{V_{\text{vol}}}\sum_{\boldsymbol{k}}{}' \langle c_{\boldsymbol{k},\uparrow}\, c_{-\boldsymbol{k},\downarrow}\rangle e^{i\boldsymbol{k}\cdot\boldsymbol{\rho}}. \tag{3.58}$$

The pair correlation function of two electrons at the same place $\boldsymbol{\rho} = 0$ is linked to the pair potential in Eq. (3.8). We note that the pair potential is the product of the pair correlation function and the attractive interaction. In the pairing Hamiltonian, the interaction is short range in real space $V(\boldsymbol{\rho}) = -g\delta(\boldsymbol{\rho})$. The pair potential satisfies the relation,

$$\Delta_{\uparrow,\downarrow} = g\langle \psi_\uparrow(\boldsymbol{r}_1)\,\psi_\downarrow(\boldsymbol{r}_1)\rangle = -g\langle \psi_\downarrow(\boldsymbol{r}_1)\,\psi_\uparrow(\boldsymbol{r}_1)\rangle = -\Delta_{\downarrow,\uparrow}. \tag{3.59}$$

Due to Pauli's exclusion principal, two electrons at the same place must have the opposite spin to each other. Above relation shows that the pair potential is anti-symmetric under the commutation of two spins. Thus, a Cooper pair belongs to spin-singlet s-wave symmetry class.

The function $F(\boldsymbol{\rho})$ describes the pairing correlation between two electrons with their relative coordinate at $\boldsymbol{\rho} \neq 0$. The correlation function is calculated at $T = 0$ as a function of $\rho = |\boldsymbol{\rho}|$,

$$F(\boldsymbol{\rho}) = \frac{1}{V_{\text{vol}}}\sum_{\boldsymbol{k}}{}' \frac{\Delta, e^{i\varphi}}{2E_{\boldsymbol{k}}} e^{i\boldsymbol{k}\cdot\boldsymbol{\rho}}. \tag{3.60}$$

By replacing the summation by integration in three dimension,

$$\frac{1}{V_{\text{vol}}}\sum_{\boldsymbol{k}}{}' = \frac{1}{(2\pi)^3}\int_0^{2\pi} d\phi \int_0^{\pi} d\theta \sin\theta \int_0^{k_0} dk k^2, \tag{3.61}$$

the correlation function is expressed as

$$F(\rho) = \frac{\Delta e^{i\varphi}}{4\pi^2} \int_0^{k_0} dk k^2 \frac{\sin(k\rho)}{k\rho} \frac{1}{\sqrt{\xi_k^2 + \Delta_0^2}}. \tag{3.62}$$

Since the integrand is large at $\xi_k = 0$, we apply the approximation $k \approx k_F + \xi/\hbar v_F$. By changing the running variable $k \to \xi$, we obtain

$$F(\rho) \approx \frac{\Delta e^{i\varphi}}{4\pi^2 \rho \hbar v_F} 2 \int_0^{\hbar\omega_D} d\xi k_F \frac{\sin(k_F\rho)\cos(\xi\rho/\hbar v_F)}{\sqrt{\xi^2 + \Delta_0^2}}, \tag{3.63}$$

$$= \frac{m k_F \Delta e^{i\varphi}}{2\pi^2 \hbar^2} \frac{\sin(k_F\rho)}{k_F\rho} K_0\left(\frac{\rho}{\pi\xi_0}\right), \quad \xi_0 \equiv \frac{\hbar v_F}{\pi \Delta_0}, \tag{3.64}$$

where K_0 is the modified Bessel function of the second kind and $K_0(x) \sim e^{-x}$ for $x \gg 1$. The characteristic length ξ_0 is called coherence length which describes the spatial size of a Cooper pair. In typical metallic superconductors, ξ_0 is about 1×10^{-6} m.

3.5 Magnetic Properties

The pair correlation function becomes non-uniform and depends on the center-of-mass coordinate $\boldsymbol{r}$ in the presence of magnetic field. We consider pairing correlation function,

$$\langle \psi_\uparrow(\boldsymbol{r})\, \psi_\downarrow(\boldsymbol{r})\rangle \propto \Psi(\boldsymbol{r}) \equiv \sqrt{n_s(\boldsymbol{r})}\, e^{i\varphi(\boldsymbol{r})}. \tag{3.65}$$

This relation defines the macroscopic wave function of superconducting condensate described by the density of Cooper pairs n_s and the phase of pairs φ. The superconducting condensate would have the kinetic energy density

$$E = \int d\boldsymbol{r}\, \frac{\hbar^2}{2m^*} \left(\nabla + i\frac{e^*}{\hbar c}\boldsymbol{A}\right) \Psi^*(\boldsymbol{r}) \cdot \left(\nabla - i\frac{e^*}{\hbar c}\boldsymbol{A}\right) \Psi(\boldsymbol{r}) \tag{3.66}$$

where $m^* = 2m$ and $e^* = 2e$ are the effective mass and the effective charge of a Cooper pair, respectively. Let us assume that the density n_s is uniform in a superconductor for a while. In the absence of the vector potential, the kinetic energy is estimated as

$$E = \int d\boldsymbol{r}\, \frac{\hbar^2 n_s}{4m} (\nabla\varphi)^2. \tag{3.67}$$

The results suggest that the spatial gradient in the phase increases the energy. Thus, the phase φ must be uniform in the ground state, (i.e., $\nabla\varphi = 0$). Equation (3.67) corresponds to the elastic energy to deform the phase from the ground state. The current density of such condensate would be described as

$$j(r) = \frac{2e\hbar}{4mi}\left[\Psi^* \left(\nabla - i\frac{2e}{\hbar c}A\right)\Psi - \left(\nabla + i\frac{2e}{\hbar c}A\right)\Psi^*\ \Psi\right], \tag{3.68}$$

$$= \frac{e\hbar}{m}n_s\left[\nabla\varphi(r) - \frac{2e}{\hbar c}A\right]. \tag{3.69}$$

The current is driven by the spatial derivative of phase, which explains the Josephson effect. The supercurrent density on the left-hand side is an observable, whereas the vector potential on the right-hand side depends on the gauge choice. The superconducting phase is necessary to be transformed as $\varphi \to \varphi + 2e/(\hbar c)\chi$ under the gauge transformation $A \to A + \nabla\chi$. Therefore,

$$\varphi(r) - \frac{2e}{\hbar c}\int^r dl \cdot A(l) \tag{3.70}$$

is referred to as a gauge-independent phase.

Meissner Effect

Let us consider a superconductor in a weak magnetic field. The magnetic field in the superconductor should be described by the Maxwell equation,

$$\nabla \times H - \frac{1}{c}\partial_t E = \frac{4\pi}{c}j. \tag{3.71}$$

In a static magnetic field, by substituting $\partial_t E = 0$, $H = \nabla \times A$, and Eq. (3.69) into the Maxwell equation, we find

$$\nabla \times H = \frac{4\pi}{c}\frac{e\hbar n_s}{m}\left(\nabla\varphi - \frac{2e}{\hbar c}A\right). \tag{3.72}$$

By multiplying $\nabla\times$ from the left and using relations

$$\nabla \times \nabla \times H = -\nabla^2 H, \quad \nabla \cdot H = 0, \quad \nabla \times (\nabla\varphi) = 0, \tag{3.73}$$

we obtain

$$-\nabla^2 H = -\frac{4\pi(2n_s)e^2}{mc^2}H. \tag{3.74}$$

We try to solve the equation at a surface of a superconductor as shown in Fig. 3.3a, where the superconductor occupies $x \geq 0$ and $x < 0$ is vacuum. A magnetic field is applied uniformly in the y-direction as $H = H_0\hat{y}$ in vacuum. A magnetic field in the

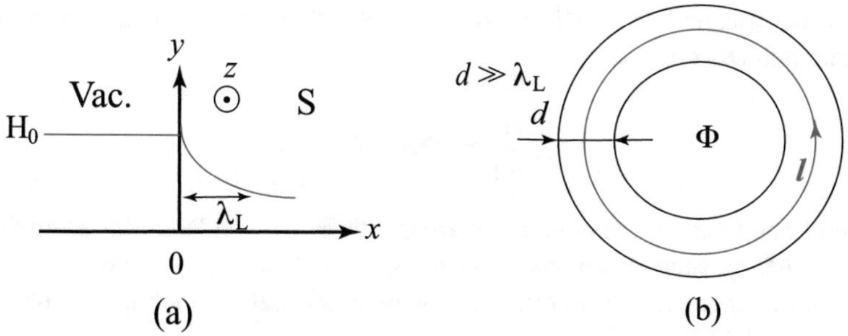

Fig. 3.3 **a** A magnetic field at a surface of a superconductor. The electric current flows in the z-direction to screen an external magnetic field. **b** The integration path $\boldsymbol{l}$ inside a superconducting ring. The arm is much thicker than the London length. The magnetic flux in the hole Φ is quantized due to the boundary condition for the macroscopic wave function

superconductor $\boldsymbol{H} = H(x)\hat{\boldsymbol{y}}$ obeys

$$\frac{d^2}{dx^2}H(x) - \frac{1}{\lambda_L^2}H(x) = 0, \quad \lambda_L \equiv \sqrt{\frac{mc^2}{4\pi n_e e^2}}, \tag{3.75}$$

where we have used a relation $2n_s = n_e$ with n_e is the electron density. The solution with a boundary condition $H(0) = H_0$,

$$H(x) = H_0 e^{-\frac{x}{\lambda_L}}, \quad \boldsymbol{j} = -\frac{c}{4\pi}\frac{H_0}{\lambda_L}e^{-\frac{x}{\lambda_L}}\hat{\boldsymbol{z}}, \tag{3.76}$$

indicates that the penetration of the magnetic field is limited in the range of λ_L and that the current flows at the surface of the superconductor to screen the applied magnetic field. The characteristic length λ_L is called London length and is about 10–1000 nm at low temperatures in typical metallic superconductors.

At the surface $0 < x < \lambda_L$, the electric current in the z-direction under the magnetic field in the y-direction feels the Lorentz force in the x-direction,

$$\boldsymbol{f} = \frac{1}{c}\boldsymbol{j} \times \boldsymbol{H} = \frac{H_0^2}{4\pi\lambda_L}e^{-\frac{2x}{\lambda_L}}\hat{\boldsymbol{x}} = -\frac{1}{4\pi}\left(\partial_x H\right)H\hat{\boldsymbol{x}}. \tag{3.77}$$

The force points to the center of the superconductor. The current, however, stays at the surface because the superconducting condensate pushes the current back. To push the current from the center ($x = \infty$) of a superconductor to its surface ($x = 0$) against the Lorentz force, the superconducting condensate does the work of

$$W = -\int_{x=\infty}^{x=0} d\boldsymbol{r} \cdot \boldsymbol{f} = \frac{1}{4\pi}\int_{\infty}^{0} dx\,\partial_x H H = \frac{H_0^2}{8\pi}\, e^{-2x/\lambda_L}\Big|_{\infty}^{0} = \frac{H_0^2}{8\pi}. \tag{3.78}$$

An external magnetic field H_0 must be smaller than the thermodynamic critical magnetic field H_c defined by

$$\frac{H_c^2}{8\pi} \equiv \Omega_N - \Omega_S. \tag{3.79}$$

The right-hand side of the equation corresponds to the condensation energy of the superconducting state and is given by $N_0\Delta_0^2/2$ at $T = 0$ as we have discussed in Sect. 3.3. For $H_0 > H_c$, the current and the magnetic fields penetrate into the superconductor and destroy superconductivity.

Flux Quantization

A superconducting ring accommodates magnetic flux Φ at its hole as shown in Fig. 3.3b. When the thickness of the arm d is much larger than λ_L, both the electric current and the magnetic field are absent at the center of the arm. Let us integrate the current in Eq. (3.69) along such a closed path along the ring,

$$0 = \oint d\boldsymbol{l} \cdot \boldsymbol{j}(\boldsymbol{l}) = \frac{e\hbar n_s}{m}\left[\oint d\boldsymbol{l} \cdot \nabla\varphi - \frac{2e}{\hbar c}\oint d\boldsymbol{l} \cdot \boldsymbol{A}\right]. \tag{3.80}$$

The second term is calculated using the Stokes theorem,

$$\frac{2e}{\hbar c}\oint d\boldsymbol{l} \cdot \boldsymbol{A} = \frac{2e}{\hbar c}\int d\boldsymbol{S} \cdot (\nabla \times \boldsymbol{A}) = \frac{2e}{\hbar c}\int d\boldsymbol{S} \cdot \boldsymbol{H} = \frac{2e}{\hbar c}\Phi. \tag{3.81}$$

The first term should satisfy

$$\frac{1}{2\pi}\oint d\boldsymbol{l} \cdot \nabla\varphi(\boldsymbol{l}) = \frac{\varphi(2\pi) - \varphi(0)}{2\pi} = n, \tag{3.82}$$

because the wave function must be single-valued in quantum mechanics. Thus, we find that the magnetic flux passing through the hole is quantized by

$$\Phi = n\phi_0, \quad \phi_0 = \frac{\pi\hbar c}{e} = 2 \times 10^{-7}\,\text{gauss} \cdot \text{cm}^2, \tag{3.83}$$

where ϕ_0 is called flux quantum. An integer number n in Eq. (3.82) is a topological invariant which is called winding number.

Vortex as a Topological Matter

Superconductors are classified into two kinds in terms of their magnetic property. Phenomenologically,

$$\kappa = \frac{\lambda_L}{\xi_0} \tag{3.84}$$

characterizes the boundary. The type I superconductor characterized by $\kappa \leq 1/\sqrt{2}$ excludes the external magnetic field up to $H < H_c$ and keeps superconductivity. Once H goes over H_c, superconductivity disappears. The threshold H_c is called the thermodynamic critical field as discussed in Eq. (3.79). On the other hand, the type II superconductor characterized by $\kappa > 1/\sqrt{2}$ have two critical fields H_{c_1} and H_{c_2}, $(H_{c_1} < H_{c_2})$. The superconductor excludes perfectly the external magnetic field in a weak magnetic field $H < H_{c_1}$. The superconductivity disappears in a strong magnetic field $H > H_{c_2}$. In the intermediate region $H_{c_1} < H < H_{c_2}$, a magnetic field penetrates into the superconductor as quantized magnetic vortices. According to textbooks, these critical fields are represented as

$$H_{c_2} = \frac{\phi_0}{2\pi\xi_0^2}, \quad H_{c_1} = \frac{\phi_0}{2\pi\lambda_{\mathrm{L}}^2}\log\left[\frac{\lambda_{\mathrm{L}}}{\xi_0}\right]. \tag{3.85}$$

Magnetic fluxes fluctuate the phase of the superconducting condensate spatially. The phase, however, must be uniform at least the spatial area of ξ_0^2 to keep superconductivity. In a high magnetic field at which a flux quanta ϕ_0 passes through the area of ξ_0^2, the condensate cannot keep the phase coherence any longer. The critical field H_{c_2} indicates such a boundary between the normal state and the superconducting state.

Let us consider a magnetic flux at the origin of two-dimensional plane. The vector potential of such a flux is represented by

$$\boldsymbol{A} = \frac{\phi}{2\pi r}\boldsymbol{e}_\theta, \tag{3.86}$$

where we used the polar coordinate in two-dimension $x = r\cos\theta$ and $y = r\sin\theta$. The magnetic field calculated as

$$H_z = \frac{1}{r}\partial_r\left(r\frac{\phi}{2\pi r}\right) = 0 \tag{3.87}$$

is zero for $r \neq 0$. On the other hand, the magnetic flux calculated as

$$\oint d\boldsymbol{l}\cdot\boldsymbol{A} = \int_0^{2\pi} d\theta r\frac{\phi}{2\pi r} = \phi \tag{3.88}$$

is finite, where $d\boldsymbol{l} = d\theta\, r\, \boldsymbol{e}_\theta$ indicates an integration path enclosing the flux at $r = 0$. Stokes theorem does not hold true because of the singularity in the vector potential at the origin. Let us assume that the magnetic flux at the origin is $\phi = n\phi_0$. In such a situation, the wave function far from the origin has the form

$$\Psi(\boldsymbol{r}) \approx \sqrt{n_s}e^{in\theta}. \tag{3.89}$$

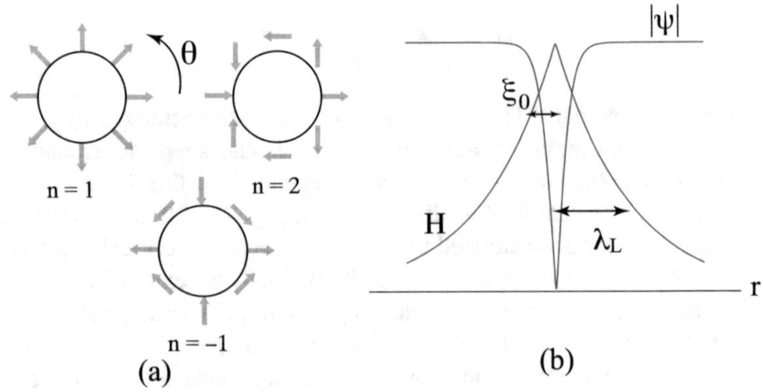

Fig. 3.4 The phase of superconducting state around a vortex core for $r \neq 0$ is illustrated in **a** for several choices of winding numbers n. The spatial profile of the macroscopic wave function and that of a magnetic field around a vortex core are shown in **b**

In Fig. 3.4a, we illustrate the phase of a superconductor far from the origin, where the phase $\varphi(\theta) = n\theta$ is indicated by arrows for $n = 1, 2$ and -1. The superconducting phase is linked to the angle in real space coordinate. At the origin $r = 0$, however, it is impossible to determine the direction of the arrows. To relax the situation, $n_s(\boldsymbol{r})$ goes to zero at $r = 0$ as shown in Fig. 3.4b. The coherence length ξ_0 characterizes the spatial variation of the wave function, whereas the London length characterizes the variation of magnetic fields. The place for $n_s(\boldsymbol{r}) = 0$ (a node in the wave function) is called vortex core. Here, let us consider a superconductor in $H_{c_1} < H < H_{c_2}$ which accommodates the magnetic flux $N\phi_0$ in total. There would be two ways for $N\phi_0$ flux to stay in the superconductor: one giant flux with the winding number $n = N$ or N flux quanta with the winding number $n = 1$ at each. According to Eq. (3.67), the energy of the former is given by $N^2 E_1$, whereas that of the latter would be $N E_1$, where $E_1 = (\hbar^2 n_s V/4m)$ is the energy density for one vortex with $n = 1$. Therefore, the magnetic fluxes pass through a superconductor as a number of flux quanta.

Finally, we briefly discuss a topological aspect of a vortex. We first consider a simple defect in a crystal lattice as shown in Fig. 3.5a, where a filled circle corresponds to an atom in the crystal. An open circle coming from the left is a vacancy. The vacancy comes into the crystal at the 1st time-step, moves to the right at 2nd and the 3rd time-step, and leaves away from the crystal. There is no difference between the crystal structures before the vacancy coming in and those after the vacancy leaving away. The vacancy is a local defect in the crystal and affects only its neighboring atoms. Next we consider a vortex core with $n = 1$ in a superconductor. Since the vortex core is a topological defect, its propagation changes the global phase configuration as illustrated in Fig. 3.5b. Before the core coming, a superconductor is in the ground state indicated by a uniform phase at $\varphi = 0$. At the 1st time-step, the superconductor changes the phase configuration drastically to accommodate the vortex. The winding number of the phase configuration becomes $n = 1$ at the second time-step. At

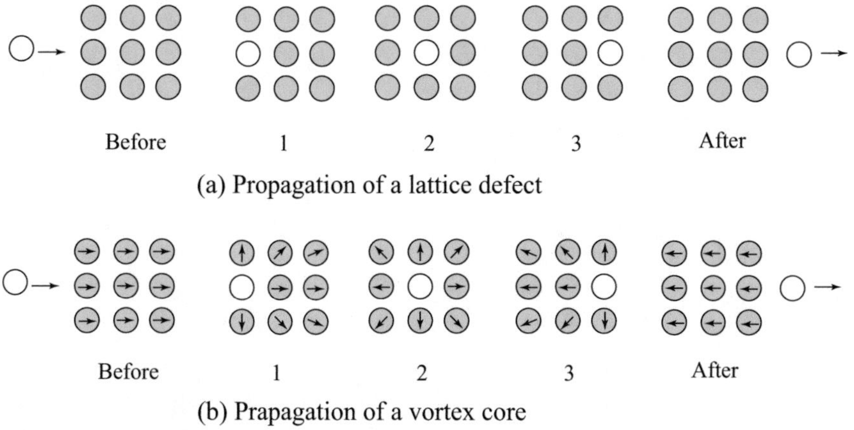

Fig. 3.5 The propagation of a defect from left to right through a uniform system. **a** A filled circle corresponds to an atom in a crystal. An open circle coming from the left is a vacancy. **b** An arrow represents a superconducting phase at the place. An open circle coming from the left indicates a magnetic flux quantum

the 3rd time-steps, the arrows tend to point to the left. After the vortex leaving, the superconductor is in the ground state with a uniform phase at $\varphi = \pi$. In the process, the winding number in the superconducting state changes as $0 \to 1 \to 0$. The propagation of a topological defect changes the phase configurations of the entire superconductor.

References

Bardeen, J., Cooper, L.N., Schrieffer, J.R.: Phys. Rev. **108**, 1175 (1957). https://doi.org/10.1103/PhysRev.108.1175

Bogoliubov, N.N.: Sov. Phys. JETP **7**, 41 (1958a)

Bogoliubov, N.N.: Sov. Phys. JETP **7**, 51 (1958b)

Chapter 4
Andreev Reflection

Abstract Andreev reflection at an interface between a superconductor and a normal metal represents the penetration of a Cooper pair into the normal metal. The Andreev reflections coefficients are calculated from the boundary conditions at the interface of the wave functions obtained by solving the Bogoliubov-de Gennes equation. By substituting the Andreev refection coefficients into the current formulas, we obtain the tunnel conductance spectra of a superconductor and Josephson current in a superconductor/insulator/superconductor junction.

4.1 Bogoliubov-de Gennes Hamiltonian

The penetration of a Cooper pair from a superconductor into a non-superconducting material modifies the electromagnetic properties of the material. Such effect is called proximity effect. In the Bogoliubov-de Gennes (BdG) picture, the Andreev reflection (Andreev 1964) represents the conversion between an electron and a Cooper pair. Since a Cooper pair is a charged particle, the propagation of a Cooper pair through a junction interface affects drastically the low energy charge transport such as the differential conductance in normal-metal/superconductor (NS) junctions and the Josephson current in superconductor/insulator/superconductor (SIS) junctions. We show schematic illustration of superconducting junctions in Fig. 4.1.

Although the pair potential is absent at a normal metal in an NS junction and at an insulator in SIS junction, a Cooper pair is present there as the pairing correlation. To analyze the electric current in such a non-superconducting segment, we begin this section with the mean-field Hamiltonian in real space,

Y. Asano, *Andreev Reflection in Superconducting Junctions*,
SpringerBriefs in Physics, https://doi.org/10.1007/978-981-16-4165-7_4

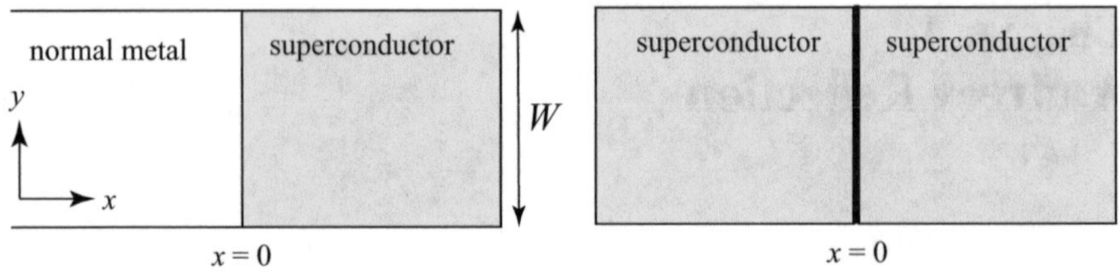

(a) NS junction (b) SIS junction

Fig. 4.1 Schematic picture of superconducting junctions are shown for an NS junction in (**a**) and an SIS junction in (**b**). The width of the junction is W. The electric current flows in the x-direction and the interface is parallel to the y-direction

$$\mathcal{H}_{\text{MF}} = \int d\boldsymbol{r} \sum_{\alpha} \psi_{\alpha}^{\dagger}(\boldsymbol{r})\, \xi(\boldsymbol{r})\, \psi_{\alpha}(\boldsymbol{r})$$

$$+ \int d\boldsymbol{r}\; \psi_{\uparrow}^{\dagger}(\boldsymbol{r})\, \Delta(\boldsymbol{r}) e^{i\varphi}\, \psi_{\downarrow}^{\dagger}(\boldsymbol{r}) + \psi_{\downarrow}(\boldsymbol{r})\, \Delta(\boldsymbol{r}) e^{-i\varphi}\, \psi_{\uparrow}(\boldsymbol{r}), \tag{4.1}$$

$$= \frac{1}{2} \int d\boldsymbol{r} [\psi_{\uparrow}^{\dagger}(\boldsymbol{r}), \psi_{\downarrow}^{\dagger}(\boldsymbol{r}), \psi_{\uparrow}(\boldsymbol{r}), \psi_{\downarrow}(\boldsymbol{r})]\; \check{H}_{\text{BdG}}(\boldsymbol{r}) \begin{bmatrix} \psi_{\uparrow}(\boldsymbol{r}) \\ \psi_{\downarrow}(\boldsymbol{r}) \\ \psi_{\uparrow}^{\dagger}(\boldsymbol{r}) \\ \psi_{\downarrow}^{\dagger}(\boldsymbol{r}) \end{bmatrix}, \tag{4.2}$$

$$\check{H}_{\text{BdG}}(\boldsymbol{r}) = \begin{bmatrix} \xi(\boldsymbol{r})\, \hat{\sigma}_0 & \Delta(\boldsymbol{r})\, e^{i\varphi}\, i\hat{\sigma}_2 \\ -\Delta(\boldsymbol{r})\, e^{-i\varphi}\, i\hat{\sigma}_2 & -\xi(\boldsymbol{r})\, \hat{\sigma}_0 \end{bmatrix}, \tag{4.3}$$

$$\xi(\boldsymbol{r}) = -\frac{\hbar^2 \nabla^2}{2m} + V(\boldsymbol{r}) - \epsilon_F, \tag{4.4}$$

where $\hat{\sigma}_j$ for $j = 1 - 3$ is Pauli's matrix in spin space and $\hat{\sigma}_0$ is the unit matrix. The derivation is shown in Appendix C. Equation (4.3) has 4×4 structure because of the spin and electron-hole degrees of freedom. The 2×2 space at the upper-left of Eq. (4.3) is the Hamiltonian of an electron (particle) and that at the lower-right is the Hamiltonian of a hole (anti-particle). The pair potential hybridizes the two spaces. The hole-degree of freedom is introduced in the BdG picture in such a way as

$$\int d\boldsymbol{r} \sum_{\alpha,\beta} \psi_{\alpha}^{\dagger}(\boldsymbol{r}) \xi_{\alpha,\beta}(\boldsymbol{r}) \psi_{\beta}(\boldsymbol{r}), \tag{4.5}$$

$$= \frac{1}{2} \int d\boldsymbol{r} \sum_{\alpha,\beta} \left[\psi_{\alpha}^{\dagger}(\boldsymbol{r})\, \xi_{\alpha,\beta}(\boldsymbol{r})\, \psi_{\beta}(\boldsymbol{r}) - \psi_{\alpha}(\boldsymbol{r})\, \xi_{\alpha,\beta}^{*}(\boldsymbol{r})\, \psi_{\beta}^{\dagger}(\boldsymbol{r}) \right]. \tag{4.6}$$

The relation holds true exactly even when ξ includes spin-dependent potentials and vector potentials. The spin structure of the pair potential is represented by $i\hat{\sigma}_2$ which means that the pair potential is antisymmetric under the permutation of two spins. The equation

$$\check{H}_{\mathrm{BdG}}(\boldsymbol{r})\begin{bmatrix} u_\nu(\boldsymbol{r})\,\hat{\sigma}_0 \\ v_\nu(\boldsymbol{r})\,(-i)\hat{\sigma}_2 \end{bmatrix} = E_\nu \begin{bmatrix} u_\nu(\boldsymbol{r})\,\hat{\sigma}_0 \\ v_\nu(\boldsymbol{r})\,(-i)\hat{\sigma}_2 \end{bmatrix} \tag{4.7}$$

is called BdG equation. It is possible to derive the similar equation

$$\check{H}_{\mathrm{BdG}}(\boldsymbol{r})\begin{bmatrix} v^*_\nu(\boldsymbol{r})(-i)\hat{\sigma}_2 \\ u^*_\nu(\boldsymbol{r})\hat{\sigma}_0 \end{bmatrix} = -E_\nu \begin{bmatrix} v^*_\nu(\boldsymbol{r})(-i)\hat{\sigma}_2 \\ u^*_\nu(\boldsymbol{r})\hat{\sigma}_0 \end{bmatrix}, \tag{4.8}$$

because of particle-hole symmetry of the Hamiltonian,

$$\begin{bmatrix} 0 & \sigma_0 \\ \sigma_0 & 0 \end{bmatrix} \check{H}^*_{\mathrm{BdG}}(\boldsymbol{r}) \begin{bmatrix} 0 & \sigma_0 \\ \sigma_0 & 0 \end{bmatrix} = -\check{H}_{\mathrm{BdG}}(\boldsymbol{r}). \tag{4.9}$$

The wave function satisfies the orthonormal property and the completeness,

$$\int d\boldsymbol{r}\ \left[u^*_\nu(\boldsymbol{r})u_\lambda(\boldsymbol{r}) + v^*_\nu(\boldsymbol{r})v_\lambda(\boldsymbol{r})\right] = \delta_{\nu,\lambda}, \tag{4.10}$$

$$\sum_\nu \begin{bmatrix} u_\nu(\boldsymbol{r})\hat{\sigma}_0 \\ v_\nu(\boldsymbol{r})(-i)\hat{\sigma}_2 \end{bmatrix} \left[u^*_\nu(\boldsymbol{r}')\hat{\sigma}_0,\, v^*_\nu(\boldsymbol{r}')i\hat{\sigma}_2\right] + \begin{bmatrix} v^*_\nu(\boldsymbol{r})(-i)\hat{\sigma}_2 \\ u^*_\nu(\boldsymbol{r})\hat{\sigma}_0 \end{bmatrix} \left[v_\nu(\boldsymbol{r}')\,i\hat{\sigma}_2,\, u_\nu(\boldsymbol{r}')\hat{\sigma}_0\right]$$
$$= \check{1}_{4\times4}\,\delta(\boldsymbol{r}-\boldsymbol{r}'). \tag{4.11}$$

The Bogoliubov transformation is given by

$$\begin{bmatrix} \psi_\uparrow(\boldsymbol{r}) \\ \psi_\downarrow(\boldsymbol{r}) \\ \psi^\dagger_\uparrow(\boldsymbol{r}) \\ \psi^\dagger_\downarrow(\boldsymbol{r}) \end{bmatrix} = \sum_\nu \begin{bmatrix} u_\nu(\boldsymbol{r})\hat{\sigma}_0 \\ v_\nu(\boldsymbol{r})(-i)\hat{\sigma}_2 \end{bmatrix} \begin{bmatrix} \gamma_{\nu,\uparrow} \\ \gamma_{\nu,\downarrow} \end{bmatrix} + \begin{bmatrix} v^*_\nu(\boldsymbol{r})(-i)\hat{\sigma}_2 \\ u^*_\nu(\boldsymbol{r})\hat{\sigma}_0 \end{bmatrix} \begin{bmatrix} \gamma^\dagger_{-\nu,\uparrow} \\ \gamma^\dagger_{-\nu,\downarrow} \end{bmatrix}, \tag{4.12}$$

$$= \sum_\nu \begin{bmatrix} u_\nu & 0 & 0 & -v^*_\nu \\ 0 & u_\nu & v^*_\nu & 0 \\ 0 & -v_\nu & u^*_\nu & 0 \\ v_\nu & 0 & 0 & u^*_\nu \end{bmatrix} \begin{bmatrix} \gamma_{\nu,\uparrow} \\ \gamma_{\nu,\downarrow} \\ \gamma^\dagger_{-\nu,\uparrow} \\ \gamma^\dagger_{-\nu,\downarrow} \end{bmatrix}. \tag{4.13}$$

4.2 Conductance in an NS Junction

The Andreev reflection occurs at an interface between a superconductor and a normal metal. In what follows, we will consider the transport property of an NS junction, where a normal metal ($x < 0$) is connected to a superconductor ($x > 0$) at $x = 0$. The BdG Hamiltonian in Eq. (4.3) is separated in two 2×2 Hamiltonian. Here, we choose the Hamiltonian for an electron with spin ↑ and a hole with spin ↓. The BdG Hamiltonian of an NS junction becomes

$$\hat{H}_{\mathrm{BdG}}(\boldsymbol{r}) = \begin{bmatrix} \xi(\boldsymbol{r}) & \Delta(\boldsymbol{r}) \\ \Delta^*(\boldsymbol{r}) & -\xi^*(\boldsymbol{r}) \end{bmatrix}, \tag{4.14}$$

with

$$\xi(\boldsymbol{r}) = -\frac{\hbar^2\nabla^2}{2m} + v_0\delta(x) - \epsilon_F, \quad \Delta(\boldsymbol{r}) = \Delta e^{i\varphi}\Theta(x). \tag{4.15}$$

In a normal metal, the BdG equation is divided into two equations,

$$\left[-\frac{\hbar^2\nabla^2}{2m} - \epsilon_F\right] u(\boldsymbol{r}) = Eu(\boldsymbol{r}) \;\; \text{and} \;\; \left[\frac{\hbar^2\nabla^2}{2m} + \epsilon_F\right] v(\boldsymbol{r}) = Ev(\boldsymbol{r}). \tag{4.16}$$

The eigenvalues and the wave functions are obtained easily. The solution of an electron and that of a hole are represented, respectively, as

$$\begin{pmatrix} 1 \\ 0 \end{pmatrix} e^{ikx}\, f_{\boldsymbol{p}}(\boldsymbol{\rho}) \;\; \text{and} \;\; \begin{pmatrix} 0 \\ 1 \end{pmatrix} e^{ikx}\, f_{\boldsymbol{p}}(\boldsymbol{\rho}). \tag{4.17}$$

Here, the first (second) wave function belongs to $E = \xi_{\boldsymbol{k}}$ ($E = -\xi_{\boldsymbol{k}}$) with $\xi_{\boldsymbol{k}} = \hbar^2(k^2 + p^2)/(2m) - \epsilon_F$. The wave function in the transverse direction to the interface is given by $f_{\boldsymbol{p}}(\boldsymbol{\rho}) = e^{i\boldsymbol{p}\cdot\boldsymbol{\rho}}/\sqrt{S}$ with S being the cross section of the junction. These dispersions are illustrated in Fig. 4.2a, where we plot the energy of an electron $\xi_{\boldsymbol{k}}$ and that of a hole $-\xi_{\boldsymbol{k}}$ for a propagation channel $\boldsymbol{p}$ as a function of the wavenumber in the x-direction k. At the Fermi level, we find that $k = k_x = \sqrt{k_F^2 - p^2}$ and that the two dispersions are degenerate because of $\xi_{k_F} = 0$. In a superconductor, two solutions given by

$$\begin{pmatrix} u_{\boldsymbol{k}}\, e^{i\varphi/2} \\ v_{\boldsymbol{k}}\, e^{-i\varphi/2} \end{pmatrix} e^{ikx}\, f_{\boldsymbol{p}}(\boldsymbol{\rho}) \;\; \text{and} \;\; \begin{pmatrix} -v_{\boldsymbol{k}}\, e^{i\varphi/2} \\ u_{\boldsymbol{k}}\, e^{-i\varphi/2} \end{pmatrix} e^{ikx}\, f_{\boldsymbol{p}}(\boldsymbol{\rho}) \tag{4.18}$$

belong to $E = E_{\boldsymbol{k}}$ and $E = -E_{\boldsymbol{k}}$, respectively. The amplitudes of wave function are given by Eq. (3.21). The dispersion of $E_{\boldsymbol{k}}$ are illustrated in Fig. 4.2b. The pair potential hybridizes the dispersion of an electron and that of a hole. As a consequence, a gap opens at the Fermi level, (i.e., $k = \pm k_x$). The dispersion for $|k| > k_x$ describes the excitation energy of an electronlike quasiparticle, where the amplitude of an electron $u_{\boldsymbol{k}}$ is larger than that of a hole $v_{\boldsymbol{k}}$. On the other hand, the dispersion for $|k| < k_x$ describes the excitation energy of a holelike quasiparticle. In this book, an electronlike (a holelike) quasiparticle is referred to as an electron (a hole) for simplicity. Equations (4.17) and (4.18) are the eigenstate vectors as a function of the wavenumber $\boldsymbol{k}$.

Next we consider a situation where an electron at the Fermi level comes into the junction from the normal metal as indicated by α in Fig. 4.2a. Due to the potential barrier at $x = 0$, the incoming wave is reflected into the electron branch as indicated by A. Simultaneously, the incoming wave is transmitted to the superconductor. Since

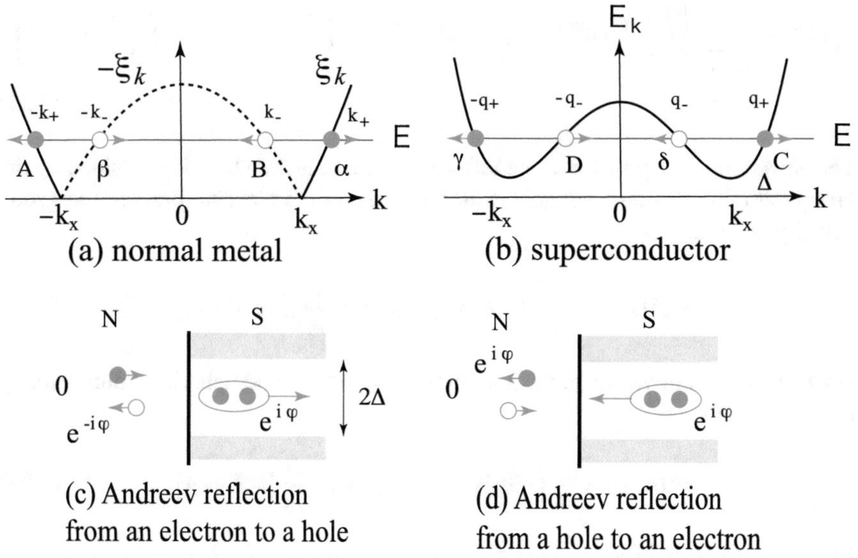

Fig. 4.2 The dispersion relation in a normal metal (**a**) and that in a superconductor (**b**). The arrow points the direction of the group velocity. The two Andreev reflection processes are schematically illustrated in (**c**) and (**d**)

the pair potential hybridizes the electron and hole branches, two outgoing waves are possible as indicated as C and D. In the normal metal, therefore, the outgoing wave from the hole channel is also possible as indicated by B. The last process is called Andreev reflection where an electron is reflected as a hole by the pair potential in a superconductor. To describe the transmission and reflection processes at the junction interface, we need the wave function for an incoming waves and those for outgoing waves at an energy E. In a normal metal, the wavenumber is calculated by the relation,

$$\pm E = \xi_k = \frac{\hbar^2 k^2}{2m} + \epsilon_p - \epsilon_F, \quad \epsilon_p = \frac{\hbar^2 p^2}{2m}. \tag{4.19}$$

The wavenumber is calculated as k_+ (k_-) in the electron (hole) branch with

$$k_\pm = \sqrt{k_x^2 \pm \frac{2mE}{\hbar^2}}, \quad k_x^2 = k_F^2 - p^2. \tag{4.20}$$

The wavenumbers at four branches in a normal metal are shown in Fig. 4.2a. The wavenumber at A becomes $-k_+$. In the hole branch, the wavenumber at B is k_- because of the relation $E = -\xi_k$. The wave function in a normal metal is represented by

$$\phi_{\mathrm{L}}(\boldsymbol{r}) = \sum_{|\boldsymbol{p}|<k_F} \left[\begin{pmatrix} 1 \\ 0 \end{pmatrix} \alpha\, e^{ik_+ x} + \begin{pmatrix} 0 \\ 1 \end{pmatrix} \beta\, e^{-ik_- x} + \begin{pmatrix} 1 \\ 0 \end{pmatrix} A\, e^{-ik_+ x} + \begin{pmatrix} 0 \\ 1 \end{pmatrix} B\, e^{ik_- x} \right] \times f_{\boldsymbol{p}}(\boldsymbol{\rho}). \tag{4.21}$$

The wavenumber is positive at the left-going hole branch, which is explained by the group velocity of a quasiparticle. Since $E = \xi_{\boldsymbol{k}}$ in the electron branch, the group velocity is defined as

$$\boldsymbol{v}^e(k, \boldsymbol{p}) = \frac{1}{\hbar} \left[\partial_k \xi_{\boldsymbol{k}}\, \hat{\boldsymbol{x}} + \partial_{\boldsymbol{p}} \xi_{\boldsymbol{k}} \right] = \frac{\hbar}{m} \left[k\, \hat{\boldsymbol{x}} + \boldsymbol{p} \right], \tag{4.22}$$

where $\hat{\boldsymbol{x}}$ is the unit vector in the x-direction. In the hole branch, the group velocity is calculated as

$$\boldsymbol{v}^h(k, \boldsymbol{p}) = \frac{1}{\hbar} \left[-\partial_k \xi_{\boldsymbol{k}}\, \hat{\boldsymbol{x}} - \partial_{\boldsymbol{p}} \xi_{\boldsymbol{k}} \right] = -\frac{\hbar}{m} \left[k\, \hat{\boldsymbol{x}} + \boldsymbol{p} \right]. \tag{4.23}$$

The group velocity of a hole is negative in the x-direction for $k > 0$. In a superconductor, the relation $E = E_k$ is transformed to

$$\xi_{\boldsymbol{k}} = \pm \Omega, \quad \Omega = \sqrt{E^2 - \Delta^2}. \tag{4.24}$$

Therefore, we obtain

$$q_\pm = \sqrt{k_x^2 \pm \frac{2m\Omega}{\hbar^2}}. \tag{4.25}$$

In the electron branch as indicated by C in Fig. 4.2b, we put $\xi_{\boldsymbol{k}} \to \Omega$ in Eq. (4.24) and the wavenumber is q_+. In the hole branch as indicated by D in Fig. 4.2b, however, we put $\xi_{\boldsymbol{k}} \to -\Omega$ in Eq. (4.24) and the wavenumber is $-q_-$. Thus, the wave function on the right-hand side of the junction interface is represented as

$$\phi_{\mathrm{R}}(\boldsymbol{r}) = \sum_{|\boldsymbol{p}|<k_F} \hat{\Phi} \left[\begin{pmatrix} u \\ v \end{pmatrix} C\, e^{iq_+ x} + \begin{pmatrix} v \\ u \end{pmatrix} D\, e^{-iq_- x} + \begin{pmatrix} u \\ v \end{pmatrix} \gamma\, e^{-iq_+ x} + \begin{pmatrix} v \\ u \end{pmatrix} \delta\, e^{iq_- x} \right] \times f_{\boldsymbol{p}}(\boldsymbol{\rho}), \tag{4.26}$$

$$u = \sqrt{\frac{1}{2}\left(1 + \frac{\Omega}{E}\right)}, \quad v = \sqrt{\frac{1}{2}\left(1 - \frac{\Omega}{E}\right)}, \quad \hat{\Phi} = \begin{pmatrix} e^{i\varphi/2} & 0 \\ 0 & e^{-i\varphi/2} \end{pmatrix}. \tag{4.27}$$

The group velocity in a superconductor is calculated as a function of $\boldsymbol{k}$,

$$\boldsymbol{v}(k, \boldsymbol{p}) = \frac{1}{\hbar} \left[\partial_k E_{\boldsymbol{k}}\, \hat{\boldsymbol{x}} + \partial_{\boldsymbol{p}} E_{\boldsymbol{k}} \right] = \frac{\hbar}{m} \left[k\, \hat{\boldsymbol{x}} + \boldsymbol{p} \right] \frac{\xi_{\boldsymbol{k}}}{E_{\boldsymbol{k}}} \tag{4.28}$$

In the electron (hole) branch, ξ_k is replaced by Ω $(-\Omega)$. The group velocities are then calculated to be

$$v^e(\pm q_+, \boldsymbol{p}) = \frac{\hbar}{m} \left[\pm q_+ \hat{\boldsymbol{x}} + \boldsymbol{p}\right] \frac{\Omega}{E}, \tag{4.29}$$

$$v^h(\pm q_-, \boldsymbol{p}) = \frac{\hbar}{m} \left[\pm q_- \hat{\boldsymbol{x}} + \boldsymbol{p}\right] \frac{-\Omega}{E}, \tag{4.30}$$

The group velocity in the x-direction is indicated by arrows in Fig. 4.2. For $E < \Delta$, the wavenumber in Eq. (4.25) become complex, which suggests that a quasiparticle at C and D in Fig. 4.2b do not propagate to $x = \infty$. In this case, all the transport channels become evanescent mode.

The boundary conditions for the wave functions are given by,

$$\phi_{\mathrm{L}}(0, y) = \phi_{\mathrm{R}}(0, y), \tag{4.31}$$

$$-\frac{\hbar^2}{2m}\left[\frac{d}{dx}\phi_{\mathrm{R}}(\boldsymbol{r})\Big|_{x=0} - \frac{d}{dx}\phi_{\mathrm{L}}(\boldsymbol{r})\Big|_{x=0}\right] + v_0\, \phi_{\mathrm{R}}(0, y) = 0. \tag{4.32}$$

The first condition represents the single-valuedness of the wave function. The second condition is derived from the BdG equation

$$\lim_{\delta\to 0} \int_{-\delta}^{\delta} dx\, H_{\mathrm{BdG}}\, \phi(\boldsymbol{r}) = E \lim_{\delta\to 0} \int_{-\delta}^{\delta} dx\, \phi(\boldsymbol{r}) = 0, \tag{4.33}$$

and implies the current conservation law. The derivation is shown in Appendix B. The wavenumber $\boldsymbol{p}$ is conserved in the transmission and reflection processes because of translational symmetry in the transverse direction. The boundary conditions are represented in a matrix form,

$$\begin{pmatrix} \alpha \\ \beta \end{pmatrix} + \begin{pmatrix} A \\ B \end{pmatrix} = \hat{\Phi}\, \hat{U} \begin{pmatrix} C \\ D \end{pmatrix} + \hat{\Phi}\, \hat{U} \begin{pmatrix} \gamma \\ \delta \end{pmatrix}, \tag{4.34}$$

$$\bar{k}\hat{\tau}_3 \begin{pmatrix} \alpha \\ \beta \end{pmatrix} - \bar{k}\hat{\tau}_3 \begin{pmatrix} A \\ B \end{pmatrix} = \hat{\Phi}\, \hat{U}\, (\bar{k}\hat{\tau}_3 + 2iz_0) \begin{pmatrix} C \\ D \end{pmatrix} - \hat{\Phi}\, \hat{U}\, (\bar{k}\hat{\tau}_3 - 2iz_0) \begin{pmatrix} \gamma \\ \delta \end{pmatrix}, \tag{4.35}$$

$$\hat{U} = \begin{pmatrix} u & v \\ v & u \end{pmatrix}, \tag{4.36}$$

where $\hat{\tau}_3$ is the third Pauli's matrix in particle-hole space. We have used the relation $\bar{k} \approx k_\pm / k_F \approx q_\pm / k_F$ which is valid at a low energy $E \ll \epsilon_F$. The transport coefficients in the normal state is calculated to be

$$t_n = \frac{\bar{k}}{\bar{k} + iz_0}, \quad r_n = \frac{-iz_0}{\bar{k} + iz_0}, \quad \bar{k} = \frac{k_x}{k_F}, \quad z_0 = \frac{m v_0}{\hbar^2 k_F}, \tag{4.37}$$

where t_n (r_n) is the transmission (reflection) coefficient of the potential barrier. The transport coefficients for a quasiparticle incoming from a normal metal can be obtained by putting $\gamma = \delta = 0$ at Eqs. (4.34) and (4.35). The coefficients relating the amplitudes of outgoing waves and those of incoming waves as

$$\begin{bmatrix} A \\ B \end{bmatrix} = \begin{bmatrix} r_{\mathrm{nn}}^{\mathrm{ee}} & r_{\mathrm{nn}}^{\mathrm{eh}} \\ r_{\mathrm{nn}}^{\mathrm{he}} & r_{\mathrm{nn}}^{\mathrm{hh}} \end{bmatrix} \begin{bmatrix} \alpha \\ \beta \end{bmatrix}, \quad \begin{bmatrix} C \\ D \end{bmatrix} = \begin{bmatrix} t_{\mathrm{sn}}^{\mathrm{ee}} & t_{\mathrm{sn}}^{\mathrm{eh}} \\ t_{\mathrm{sn}}^{\mathrm{he}} & t_{\mathrm{sn}}^{\mathrm{hh}} \end{bmatrix} \begin{bmatrix} \alpha \\ \beta \end{bmatrix} \tag{4.38}$$

are calculated to be

$$r_{\mathrm{nn}}^{\mathrm{ee}} = \frac{2r_n\Omega}{\Xi_{\mathrm{NS}}}, \quad r_{\mathrm{nn}}^{\mathrm{hh}} = \frac{2r_n^*\Omega}{\Xi_{\mathrm{NS}}}, \quad r_{\mathrm{nn}}^{\mathrm{he}} = \frac{|t_n|^2\Delta e^{-i\varphi}}{\Xi_{\mathrm{NS}}}, \quad r_{\mathrm{nn}}^{\mathrm{eh}} = \frac{|t_n|^2\Delta e^{i\varphi}}{\Xi_{\mathrm{NS}}}, \tag{4.39}$$

$$t_{\mathrm{sn}}^{\mathrm{ee}} = \frac{t_n\, e^{-i\varphi/2}\sqrt{2E}\sqrt{E+\Omega}}{\Xi_{\mathrm{NS}}}, \quad t_{\mathrm{sn}}^{\mathrm{hh}} = \frac{t_n^*\, e^{i\varphi/2}\sqrt{2E}\sqrt{E+\Omega}}{\Xi_{\mathrm{NS}}}, \tag{4.40}$$

$$t_{\mathrm{sn}}^{\mathrm{he}} = \frac{r_n^* t_n\, e^{-i\varphi/2}\sqrt{2E}\sqrt{E-\Omega}}{\Xi_{\mathrm{NS}}}, \quad t_{\mathrm{sn}}^{\mathrm{eh}} = \frac{r_n t_n^*\, e^{i\varphi/2}\sqrt{2E}\sqrt{E-\Omega}}{\Xi_{\mathrm{NS}}}, \tag{4.41}$$

$$\Xi_{\mathrm{NS}} = 2\Omega + |t_n|^2(E-\Omega). \tag{4.42}$$

By putting $\alpha = \beta = 0$, the transport coefficients for a quasiparticle incoming from a superconductor become

$$\begin{bmatrix} C \\ D \end{bmatrix} = \begin{bmatrix} r_{\mathrm{ss}}^{\mathrm{ee}} & r_{\mathrm{ss}}^{\mathrm{eh}} \\ r_{\mathrm{ss}}^{\mathrm{he}} & r_{\mathrm{ss}}^{\mathrm{hh}} \end{bmatrix} \begin{bmatrix} \gamma \\ \delta \end{bmatrix}, \quad \begin{bmatrix} A \\ B \end{bmatrix} = \begin{bmatrix} t_{\mathrm{ns}}^{\mathrm{ee}} & t_{\mathrm{ns}}^{\mathrm{eh}} \\ t_{\mathrm{ns}}^{\mathrm{he}} & t_{\mathrm{ns}}^{\mathrm{hh}} \end{bmatrix} \begin{bmatrix} \gamma \\ \delta \end{bmatrix} \tag{4.43}$$

with

$$r_{\mathrm{ss}}^{\mathrm{ee}} = \frac{2r_n\Omega}{\Xi_{\mathrm{NS}}}, \quad r_{\mathrm{ss}}^{\mathrm{hh}} = \frac{2r_n^*\Omega}{\Xi_{\mathrm{NS}}}, \quad r_{\mathrm{ss}}^{\mathrm{he}} = \frac{-|t_n|^2\Delta}{\Xi_{\mathrm{NS}}}, \quad r_{\mathrm{ss}}^{\mathrm{eh}} = \frac{-|t_n|^2\Delta}{\Xi_{\mathrm{NS}}}, \tag{4.44}$$

$$t_{\mathrm{ns}}^{\mathrm{ee}} = \frac{2t_n\Omega e^{i\varphi/2}}{\Xi_{\mathrm{NS}}}\sqrt{\frac{E+\Omega}{2E}}, \quad t_{\mathrm{ns}}^{\mathrm{hh}} = \frac{2t_n^*\Omega e^{-i\varphi/2}}{\Xi_{\mathrm{NS}}}\sqrt{\frac{E+\Omega}{2E}} \tag{4.45}$$

$$t_{\mathrm{ns}}^{\mathrm{he}} = \frac{2t_n^* r_n e^{-i\varphi/2}\Omega}{\Xi_{\mathrm{NS}}}\sqrt{\frac{E-\Omega}{2E}}, \quad t_{\mathrm{ns}}^{\mathrm{eh}} = \frac{2t_n r_n^* e^{i\varphi/2}\Omega}{\Xi_{\mathrm{NS}}}\sqrt{\frac{E-\Omega}{2E}}. \tag{4.46}$$

The scattering matrix of the junction interface is then represented by

$$\begin{bmatrix} A \\ B \\ C \\ D \end{bmatrix} = \mathbb{S} \begin{bmatrix} \alpha \\ \beta \\ \gamma \\ \delta \end{bmatrix}, \quad \mathbb{S} = \begin{bmatrix} r_{\mathrm{nn}}^{\mathrm{ee}} & r_{\mathrm{nn}}^{\mathrm{eh}} & \tilde{t}_{\mathrm{ns}}^{\mathrm{ee}} & \tilde{t}_{\mathrm{ns}}^{\mathrm{eh}} \\ r_{\mathrm{nn}}^{\mathrm{he}} & r_{\mathrm{nn}}^{\mathrm{hh}} & \tilde{t}_{\mathrm{ns}}^{\mathrm{he}} & \tilde{t}_{\mathrm{ns}}^{\mathrm{hh}} \\ \tilde{t}_{\mathrm{sn}}^{\mathrm{ee}} & \tilde{t}_{\mathrm{sn}}^{\mathrm{eh}} & r_{\mathrm{ss}}^{\mathrm{ee}} & r_{\mathrm{ss}}^{\mathrm{eh}} \\ \tilde{t}_{\mathrm{sn}}^{\mathrm{he}} & \tilde{t}_{\mathrm{sn}}^{\mathrm{hh}} & r_{\mathrm{ss}}^{\mathrm{he}} & r_{\mathrm{ss}}^{\mathrm{hh}} \end{bmatrix}. \tag{4.47}$$

The transmission coefficients are corrected by the ratio of the velocity in the outgoing channel and that in the incoming channel, (i.e., $\sqrt{v_{\mathrm{out}}/v_{\mathrm{in}}}$). The results are given by

$$\tilde{t}^{ij}_{\rm sn} = \sqrt{{\rm Re}\left[\frac{\Omega}{E}\right]}t^{ij}_{\rm sn}, \quad \tilde{t}^{ij}_{\rm ns} = \sqrt{{\rm Re}\left[\frac{E}{\Omega}\right]}t^{ij}_{\rm ns}. \tag{4.48}$$

For $E > \Delta$, all the channels in a superconductor are propagating with the velocity of $|v_S| = v_{k_x}(\Omega/E)$. In a normal metal, all the channels are always propagating with the velocity of $|v_N| = v_{k_x}$. For $E < \Delta$, on the other hand, all the transport channels in a superconductor are evanescent, which results in $\tilde{t}^{ij}_{\rm sn} = 0$ and $\tilde{t}^{ij}_{\rm ns} = 0$. The unitarity of the scattering matrix $\mathbb{S}\,\mathbb{S}^\dagger = 1$ implies the current conservation law.

The Andreev reflection of a quasiparticle is a peculiar reflection process by a superconductor. When an energy of an incident electron is smaller than the gap $E < \Delta$, an electron cannot go into a superconductor because of a gap at the Fermi level. In the superconducting gap, however, a number of Cooper pairs condense coherently at a phase of φ. A Cooper pair is a composite particle consisting of two electrons and a phase as shown in Fig. 4.2c. An incident electron has to find a partner of a pair and get the phase of $e^{i\varphi}$ so that the electron can penetrate into the superconductor as a Cooper pair. As a result of such Cooper pairing, the empty shell of the partner is reflected as a hole into the normal metal. Simultaneously, the conjugate phase $e^{-i\varphi}$ is copied to the wave function of a hole. Thus, an electron is converted to a Cooper pair at the Andreev reflection. The inverse process, the reflection from a hole to an electron, is also called the Andreev reflection as shown in Fig. 4.2d. A hole cancels an electron of a Cooper pair. As a consequence, a remaining electron is reflected into the normal metal with the phase of $e^{i\varphi}$. At $E = 0$, the wavenumber of an electron is calculated as

$$q_+ = k_F\left(1 + \frac{i}{2}\frac{\Delta}{\epsilon_F}\right) = k_F + i\frac{\Delta}{\hbar v_F}, \tag{4.49}$$

where we choose $\boldsymbol{p} = 0$ for simplicity. The inverse of the imaginary part characterizes the penetration length of an electron and is equal to the coherence length ξ_0. Namely, an incident electron decreases its amplitude and condenses as a Cooper pair in the superconducting gap. Such event happens at an interface of a superconductor and a normal metal. At $E < \Delta$, the conservation law

$$|r^{\rm ee}_{\rm nn}|^2 + |r^{\rm he}_{\rm nn}|^2 = 1 \tag{4.50}$$

holds true as there is no propagating channel in a superconductor. When the normal transmission probability is much smaller than unity $|t_n|^2 \ll 1$, the Andreev reflection is suppressed $|r^{\rm he}_{\rm nn}| \to 0$ and the normal reflection is perfect $|r^{\rm ee}_{\rm nn}| \to 1$. On the other hand at $t_n = 1$, the normal reflection is absent $|r^{\rm ee}_{\rm nn}| = 0$ and the Andreev reflection is perfect $|r^{\rm he}_{\rm nn}| \to 1$. The reflection coefficients

$$r^{\rm he}_{\rm nn} = e^{-i\arctan\left(\sqrt{\Delta^2-E^2}/E\right)}e^{-i\varphi}, \quad r^{\rm eh}_{\rm nn} = e^{-i\arctan\left(\sqrt{\Delta^2-E^2}/E\right)}e^{i\varphi} \tag{4.51}$$

includes only the phase information at $t_n = 1$.

The Blonder-Tinkham-Klapwijk (BTK) formula (Blonder et al. 1982) or Takane-Ebisawa formula (Takane and Ebisawa 1992) enables us to calculate the differential conductance of an NS junction by using the reflection coefficients,

$$G_{\mathrm{NS}} = \left.\frac{dI}{dV}\right|_{eV} = \frac{2e^2}{h}\sum_p \left(1 - |r_{\mathrm{nn}}^{\mathrm{ee}}|^2 + |r_{\mathrm{nn}}^{\mathrm{he}}|^2\right)\Big|_{E=eV}. \tag{4.52}$$

In the normal state, the Andreev reflection is absent $r_{\mathrm{nn}}^{\mathrm{he}} = 0$ and $r_{\mathrm{nn}}^{\mathrm{ee}}$ is replaced by r_n in Eq. (4.37). In such case, the formula is identical to Landauer's conductance formula by using the conservation law $|t_n|^2 = 1 - |r_n|^2$. In Appendix B, a derivation of Landauer's conductance formula is explained briefly. The formula in Eq. (4.52) also assumes the reservoirs of an electron at $x = \pm\infty$. The difference in the chemical potentials in these reservoirs corresponds to the bias voltage across the junction $eV = \mu_N - \mu_S$, where μ_N (μ_S) is the chemical potential of the reservoir connected with a normal metal (a superconductor). The voltage drops at the insulating barrier at the junction interface. The Andreev reflection describes the effects of the pair potential in a superconductor on the electric current through an NS junction. By substituting the reflection coefficients, we obtain

$$G_{\mathrm{NS}} = \frac{2e^2}{h}\sum_p \left.\frac{2\Delta^2|t_n|^4}{(2-|t_n|^2)^2(\Delta^2 - E^2) + |t_n|^4 E^2}\right|_{E=eV} \quad \text{for} \quad 0 < eV < \Delta, \tag{4.53}$$

and

$$G_{\mathrm{NS}} = \frac{2e^2}{h}\sum_p \left.\frac{2|t_n|^2[|t_n|^2E^2 + (2-|t_n|^2)\sqrt{E^2-\Delta^2}E]}{[(2-|t_n|^2)\sqrt{E^2-\Delta^2} + |t_n|^2 E]^2}\right|_{E=eV} \quad \text{for} \quad eV \geq \Delta. \tag{4.54}$$

We plot the conductance as a function of the bias voltage in Fig. 4.3, where the vertical axis is normalized to the conductance of the junction in the normal state,

$$G_{\mathrm{N}} = \frac{2e^2}{h}\sum_p |t_n|^2 = R_{\mathrm{N}}^{-1}. \tag{4.55}$$

At $|t_n| = 1$, conductance becomes

$$G_{\mathrm{NS}} = G_{\mathrm{N}} \times \begin{cases} 2 & : 0 \leq eV < \Delta \\ \frac{eV}{eV+\sqrt{(eV)^2-\Delta^2}} & : \Delta \leq eV \end{cases}, \quad G_{\mathrm{N}} = \frac{2e^2}{h}N_c, \tag{4.56}$$

where N_c is the number of propagating channels on the Fermi surface in the normal state. The conductance remains finite even though any scatterers are absent in a

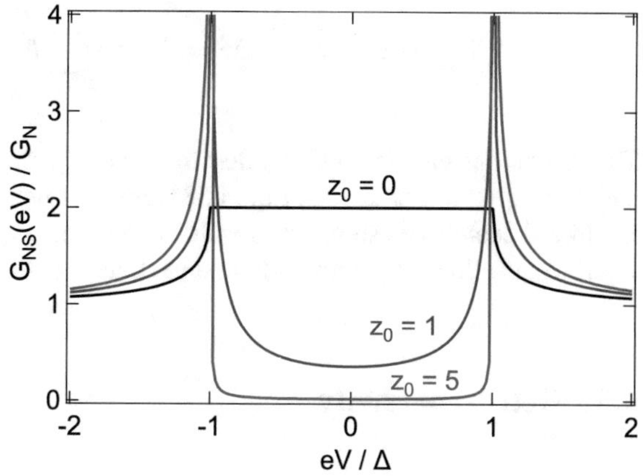

Fig. 4.3 The differential conductance of a NS junction is plotted as a function of the bias voltage across the junction for several choices of the potential barrier height z_0

sample. As discussed briefly in Appendix B, the normal conductance $G_{\rm N}$ in Eq. (4.56) due to the finite width of a sample is called Sharvin conductance. The Sharvin resistance is the inverse of the Sharvin conductance. The conductance $G_{\rm NS}$ is twice its normal value at the subgap regime $eV < \Delta$ due to the perfect Andreev reflection. An incident electron is completely converted to a Cooper pair, doubling the number of the charges passing through the interface. For large bias voltage $eV \gg \Delta$, $G_{\rm NS}$ at $t_n = 1$ goes to $G_{\rm N}$ because the Andreev reflection does not happen in such a high energy region. On the other hand in the limit of $|t_n| \ll 1$, the conductance becomes

$$G_{\rm NS} = \frac{2e^2}{h} \sum_p |t_n|^4 \frac{\Delta^2}{2(\Delta^2 - (eV)^2)} \ll G_{\rm N}, \tag{4.57}$$

for subgap regime $0 \le eV < \Delta$. The results above the gap $\Delta \le eV$

$$G_{\rm NS} = G_{\rm N} \left. \frac{E}{\sqrt{E^2 - \Delta^2}} \right|_{E=eV} \tag{4.58}$$

is proportional to the density of states in a uniform superconductor. The results in Fig. 4.3 at $z_0 = 5$ correspond to the spectra measured by scanning tunneling microscopy (STM) or scanning tunneling spectroscopy (STS) experiments.

At the end of this subsection, we explain the relation between the property of Ω and the causality of the transport coefficients. Outgoing waves from the junction are related to an incoming wave to the junction by the transmission and reflection coefficients in Eq. (4.38). In such case, the transport coefficients belong the retarded causality. The analytic properties of the retarded function $\Omega^R \equiv \sqrt{(E + i\delta)^2 - \Delta^2}$ and the advanced function $\Omega^A \equiv \sqrt{(E - i\delta)^2 - \Delta^2}$ are described by

$$\lim_{\delta\to 0^+} \sqrt{(E \pm i\delta)^2 - \Delta^2} = \begin{cases} \pm\sqrt{E^2 - \Delta^2} & E \geq \Delta \\ i\sqrt{\Delta^2 - E^2} & -\Delta < E < \Delta \\ \mp\sqrt{E^2 - \Delta^2} & E \leq -\Delta. \end{cases} \tag{4.59}$$

These relations enable a unified description of the transport coefficients for $E > 0$ and those $E < 0$. For $E < 0$, Eq. (4.53) remains unchanged, whereas $\sqrt{E^2 - \Delta^2}$ in Eq. (4.54) changes its sign. As a result, we find that Eq. (4.54) is an even function of E. In what follows, we implicitly assume that Ω belongs to the retarded causality.

4.3 Retroreflectivity

The Andreev reflection exhibits a unique property also in real space. When the wavenumber of an incident electron is $(k, \boldsymbol{p})$, the wavenumber at outgoing waves are summarized as

$$A : (-k, \boldsymbol{p}), \quad B : (k, \boldsymbol{p}), \quad C : (k, \boldsymbol{p}), \quad D : (-k, \boldsymbol{p}). \tag{4.60}$$

The trajectories of outgoing waves are illustrated in Fig. 4.4a, where we consider the group velocities in Eqs. (4.22), (4.23), (4.29), and (4.30). The velocity component perpendicular to the interface changes sign in the normal reflection, whereas all the velocity components change signs in the Andreev reflection. As a result, a reflected hole traces back the original trajectory of an incoming electron as shown in Fig. 4.4a. Such property of a hole in the Andreev reflection is called retroreflectivity. A hole of the Andreev reflection at the Fermi level ($E = 0$) is retroreflective exactly in the presence of time-reversal symmetry. Therefore, the two trajectories are retroreflective to each other even in the presence of impurities as shown in Fig. 4.4b. Although the pair potential is zero in a normal metal, a Cooper penetrates from a superconductor. A Cooper in a normal metal is described by a retroreflective electron-hole pair in the BdG picture. This fact is very important to understand the Josephson effect through a dirty metal. In the absence of time-reversal symmetry, the retroreflectivity is lost. A good example might be the motion of a quasiparticle in an external magnetic field perpendicular to the two-dimensional plane as shown in Fig. 4.4c. The Lorentz force acts on a charged particle moving in a magnetic field

$$\boldsymbol{f} = e\boldsymbol{v} \times \boldsymbol{H}. \tag{4.61}$$

The charge of a hole has the opposite sign to that in the electron. Simultaneously, the velocity of a hole is just the opposite to that of an electron. Thus, the Lorentz force acts in the same direction in the two particle branches. In Fig. 4.4c, an electron and a hole depart at a same point on the NS interface and meet again at an another same point on the interface. Such correlated motion of an electron and a hole is responsible

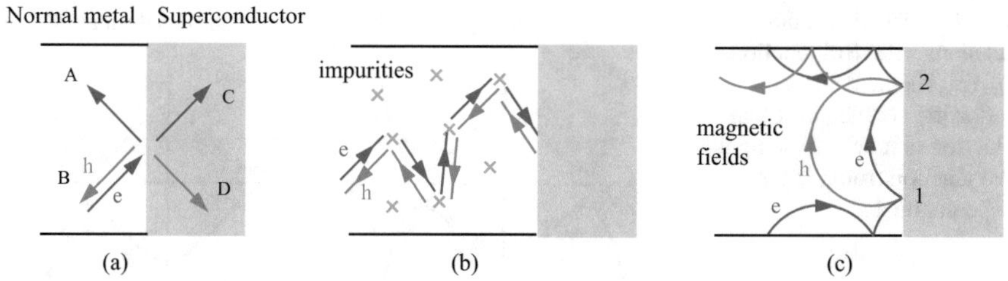

Fig. 4.4 The trajectory of outgoing waves. The Andreev reflection is retroreflective on the Fermi surface in the presence of time-reversal symmetry (**a**). A hole traces back the trajectory of an incoming electron as shown in (**a**) and (**b**). In a magnetic field, the retroreflectivity is lost as shown in (**c**)

for the Aharonov-Bohm like effect in the magnetoconductance of a ballistic NS junction (Asano 2000).

4.4 Josephson Current in an SIS Junction

The Josephson effect is a phase-coherent transport phenomenon in superconducting junctions. The dissipationless electric current flows between two superconductors at zero bias voltage. In this section, we try to understand the Josephson effect in terms of the Andreev reflection. A typical example of Josephson junction is an SIS junction as shown in Fig. 4.1b, where two superconductors sandwich an insulating barrier. The Josephson current is calculated based on Furusaki-Tsukada formula (Furusaki and Tsukada 1991) ,

$$J = \frac{e}{\hbar} \sum_{\boldsymbol{p}} k_B T \sum_{\omega_n} \frac{\Delta}{\Omega_n} \left[r_{\mathrm{he}}(\boldsymbol{p}, \omega_n) - r_{\mathrm{eh}}(\boldsymbol{p}, \omega_n) \right], \tag{4.62}$$

where the current is expressed in terms of two Andreev reflection coefficients r_{he} and r_{eh}.

The reflection coefficient r_{he} represents the process where an electron incoming from the left superconductor is reflected as a hole after traveling whole the junction as shown in Fig. 4.5. This process carries a Cooper pair from the left superconductor to the right superconductor. The coefficient r_{eh} represents the conjugate process to r_{he}. The formula suggests that the net current should be described by the subtraction of the two coefficients. Since the direct-current Josephson effect happens in equilibrium, the reflection coefficients are described as a function of a Matsubara frequency $\omega_n = (2n+1)\pi k_B T/\hbar$, where n is an integer number and T is a temperature. The analytic continuation in energy is summarized as

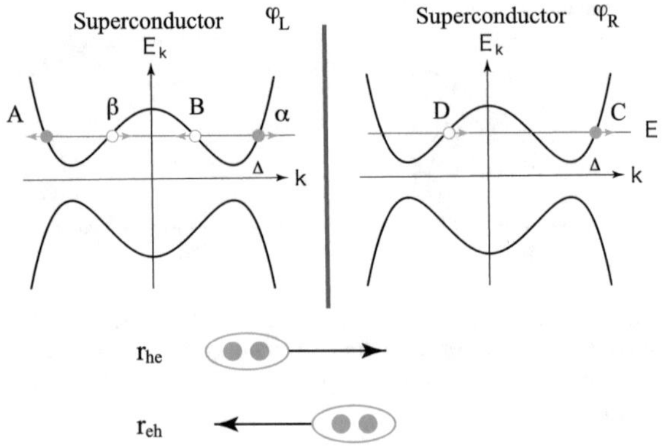

Fig. 4.5 The dispersion relations in an SIS junction. r_{he} (r_{eh}) is the Andreev reflection coefficients of an electron (a hole) incoming to the junction from the left superconductor

$$E \to i\hbar\omega_n, \quad \Omega = \sqrt{E^2 - \Delta^2} \to i\Omega_n, \quad \Omega_n = \sqrt{\hbar^2\omega_n^2 + \Delta^2}. \tag{4.63}$$

The conductance in an NS junction is represented by the absolute values of reflection coefficients in Eq. (4.52). The Josephson current in Eq. (4.62) is represented using reflection coefficients themselves. The two superconductors exchange their phase information through the phase degree of freedom of the transport coefficients. Landauer's conductance formula is a basic formula in mesoscopic physics. The formula in Eq. (4.62) as well as Eq. (4.52) bridges the physics of superconductivity and physics of mesoscopic transport phenomena.

Our next task is to calculate the Andreev reflection coefficients in an SIS junction by solving the BdG equation,

$$\begin{bmatrix} \xi(\boldsymbol{r}) & \Delta(\boldsymbol{r}) \\ \Delta(\boldsymbol{r})^* & -\xi(\boldsymbol{r}) \end{bmatrix} \begin{bmatrix} u(\boldsymbol{r}) \\ v(\boldsymbol{r}) \end{bmatrix} = E \begin{bmatrix} u(\boldsymbol{r}) \\ v(\boldsymbol{r}) \end{bmatrix}, \tag{4.64}$$

$$\Delta(\boldsymbol{r}) = \begin{cases} \Delta e^{i\varphi_L} & x < 0 \\ \Delta e^{i\varphi_R} & x > 0. \end{cases} \tag{4.65}$$

The wave function in the left superconductor is given by

$$\phi_{\mathrm{L}}(\boldsymbol{r}) = \hat{\Phi}_L \left[\begin{pmatrix} u \\ v \end{pmatrix} e^{ik_x x}\alpha + \begin{pmatrix} v \\ u \end{pmatrix} e^{-ik_x x}\beta + \begin{pmatrix} u \\ v \end{pmatrix} e^{-ik_x x} A + \begin{pmatrix} v \\ u \end{pmatrix} e^{ik_x x} B \right] \times f_{\boldsymbol{p}}(\boldsymbol{\rho}), \tag{4.66}$$

$$\hat{\Phi}_j = \begin{pmatrix} e^{i\varphi_j/2} & 0 \\ 0 & e^{-i\varphi_j/2} \end{pmatrix}, \quad u = \sqrt{\frac{1}{2}\left(1 + \frac{\Omega_n}{\hbar\omega_n}\right)}, \quad v = \sqrt{\frac{1}{2}\left(1 - \frac{\Omega_n}{\hbar\omega_n}\right)}, \tag{4.67}$$

where A and B are the amplitude of outgoing waves and α and β are the amplitude of incoming waves. In the right superconductor, the wave function is represented as

$$\phi_{\mathrm{R}}(\boldsymbol{r}) = \hat{\Phi}_R \left[\begin{pmatrix} u \\ v \end{pmatrix} e^{ik_p x} C + \begin{pmatrix} v \\ u \end{pmatrix} e^{-ik_p x} D + \begin{pmatrix} u \\ v \end{pmatrix} e^{-ik_p x} \gamma + \begin{pmatrix} v \\ u \end{pmatrix} e^{ik_p x} \delta \right] \times f_p(\boldsymbol{\rho}). \tag{4.68}$$

By substitute the wave functions into the boundary conditions in Eqs. (4.31) and (4.32), we obtain the relationship among the amplitudes of wave functions,

$$\hat{\Phi}\hat{U}\left[\begin{pmatrix} \alpha \\ \beta \end{pmatrix} + \begin{pmatrix} A \\ B \end{pmatrix}\right] = \hat{U}\begin{pmatrix} C \\ D \end{pmatrix} + \hat{U}\begin{pmatrix} \gamma \\ \delta \end{pmatrix},$$

$$\hat{\Phi}\hat{U}\left[\bar{k}\hat{\tau}_3\begin{pmatrix} \alpha \\ \beta \end{pmatrix} - \bar{k}\hat{\tau}_3\begin{pmatrix} A \\ B \end{pmatrix}\right] = \left[\hat{U}\bar{k}\hat{\tau}_3 + 2iz_0\hat{U}\right]\begin{pmatrix} C \\ D \end{pmatrix} - \left[\hat{U}\bar{k}\hat{\tau}_3 - 2iz_0\hat{U}\right]\begin{pmatrix} \gamma \\ \delta \end{pmatrix},$$

$$\hat{\Phi} = \begin{pmatrix} e^{i(\varphi_L - \varphi_R)/2} & 0 \\ 0 & e^{-i(\varphi_L - \varphi_R)/2} \end{pmatrix}, \quad \hat{U} = \begin{pmatrix} u & v \\ v & u \end{pmatrix}.$$

By eliminating C and D under $\gamma = \delta = 0$, we obtain

$$\begin{pmatrix} A \\ B \end{pmatrix} = \begin{pmatrix} r_{\mathrm{ee}} & r_{\mathrm{eh}} \\ r_{\mathrm{he}} & r_{\mathrm{hh}} \end{pmatrix} \begin{pmatrix} \alpha \\ \beta \end{pmatrix}, \tag{4.69}$$

$$r_{\mathrm{he}} = \frac{|t_n|^2 \Delta}{2i\, \Xi_{\mathrm{SS}}} \left[\hbar\omega_n(\cos\varphi - 1) + i\Omega_n \sin\varphi\right], \tag{4.70}$$

$$r_{\mathrm{eh}} = \frac{|t_n|^2 \Delta}{2i\, \Xi_{\mathrm{SS}}} \left[\hbar\omega_n(\cos\varphi - 1) - i\Omega_n \sin\varphi\right], \tag{4.71}$$

$$\Xi_{\mathrm{SS}} = (\hbar\omega_n)^2 + |t_n|^2\Delta^2\left[1 - |t_n|^2 \sin^2\left(\frac{\varphi}{2}\right)\right], \quad \varphi = \varphi_L - \varphi_R, \tag{4.72}$$

where we have used the relations

$$uv = \frac{\Delta}{2i\hbar\omega_n}, \quad u^2 - v^2 = \frac{\sqrt{(\hbar\omega_n)^2 + \Delta^2}}{\hbar\omega_n}. \tag{4.73}$$

The Josephson current is represented as

$$J = \frac{e}{\hbar}\sin\varphi \sum_p k_B T \sum_{\omega_n} \frac{|t_n|^2\Delta^2}{\hbar^2\omega_n^2 + \Delta^2\left(1 - |t_n|^2\sin^2\left(\frac{\varphi}{2}\right)\right)}. \tag{4.74}$$

The identity

$$k_B T \sum_{\omega_n} \frac{1}{(\hbar\omega_n)^2 + y^2} = \frac{1}{2y}\tanh\left(\frac{y}{2k_B T}\right) \tag{4.75}$$

enables us to carry out the summation over Matsubara frequency. Finally, we reach at

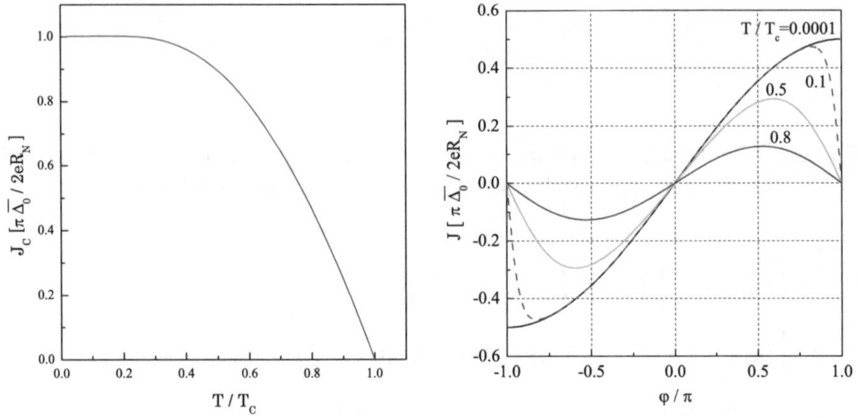

Fig. 4.6 Left: the amplitude of the Josephson current in a SIS junction ($|t_n| \ll 1$) is plotted as a function of temperature. Right: the current-phase relationship in a S-constriction-S junction ($|t_n| = 1$)

$$J = \frac{e}{\hbar} \sin\varphi \sum_p \frac{|t_n|^2 \Delta}{2\sqrt{1 - |t_n|^2 \sin^2\left(\frac{\varphi}{2}\right)}} \tanh\left(\frac{\Delta\sqrt{1 - |t_n|^2 \sin^2\left(\frac{\varphi}{2}\right)}}{2k_B T}\right). \tag{4.76}$$

First we consider the tunneling limit of $|t_n| \ll 1$ which represents an SIS junction. Using the normal resistance in Eq. (4.55), the Josephson current becomes

$$J = \frac{\pi \Delta_0}{2eR_\mathrm{N}} \left(\frac{\Delta}{\Delta_0}\right) \tanh\left(\frac{\Delta}{2k_B T}\right) \sin\varphi, \tag{4.77}$$

where Δ_0 is the amplitude of pair potential at zero temperature. Equation (4.77) was first derived by Ambegaokar-Baratoff (1963) and has explained well the Josephson effect observed in experiments. By using the dependence of Δ on temperature in Fig. 3.2, the amplitude of the Josephson current is plotted as a function of temperature in Fig. 4.6. The Josephson current increases with the decrease of temperature and saturates at low temperatures.

Next we consider a junction through a constriction by choosing $t_n = 1$. Using the Sharvin resistance R_N in Eq. (4.56), the Josephson current becomes

$$J = \frac{\pi \Delta_0}{eR_\mathrm{N}} \left(\frac{\Delta}{\Delta_0}\right) \sin\left(\frac{\varphi}{2}\right) \tanh\left(\frac{\Delta \cos\left(\frac{\varphi}{2}\right)}{2k_B T}\right). \tag{4.78}$$

This formula was first derived by Kulik-O'melyanchuck (1977). The results at $T = 0$ indicate that the current-phase relationship becomes fractional as $\sin(\varphi/2)$. The Josephson current is discontinuous at $\varphi = \pm\pi$. At a temperature near T_c, the current-phase relationship is sinusoidal as displayed in Fig. 4.6. Generally speaking, the

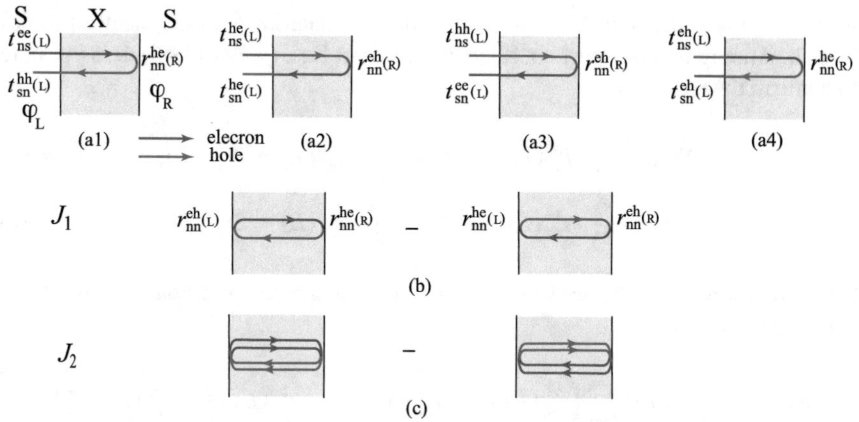

Fig. 4.7 **a** The process include the Andreev reflection at the right interface once. The material at the central segment is indicated by X. The transmission coefficient of X are t_X in the electron branch and t_X^* in the hole branch. **b** The reflection processes contribute to J_1. **c** The schematic illustration of the Andreev reflection processes in J_2.

Josephson current is decomposed into a series of

$$J = \sum_{n=1} J_n \sin(n\varphi), \tag{4.79}$$

when the junction preserves time-reversal symmetry at $\varphi = 0$. The current-phase relationship in Eq. (4.78) is described as

$$\sin\left(\frac{\varphi}{2}\right) = \sum_{n=1}^{\infty} \frac{(-1)^{n+1} 8n}{\pi(2n+1)(2n-1)} \sin(n\varphi). \tag{4.80}$$

Namely, not only the lowest order coupling term J_1 but also higher order terms J_n for $n = 2, 3, \cdots$ contribute to the Josephson current.

In what follows, we analyze the Andreev reflection processes for the lowest order term as shown in Fig. 4.7a, where a quasiparticle suffers the Andreev reflection at the right interface once. The two processes are possible for r^{he} in Eq. (4.62) as shown in (a1) and (a2). Figures in (a3) and (a4) represent the lowest order processes for r^{eh}. The contributions of these four processes to the Josephson current are summarized as

$$\begin{aligned} A_1 =& \frac{e}{\hbar} \sum_p k_B T \sum_{\omega_n} \frac{\Delta}{\Omega_n} \Big[t^{\text{hh}}_{\text{sn}}(L) \cdot t_X^* \cdot r^{\text{he}}_{\text{nn}}(R) \cdot t_X \cdot t^{\text{ee}}_{\text{ns}}(L) + t^{\text{he}}_{\text{sn}}(L) \cdot t_X \cdot r^{\text{eh}}_{\text{nn}}(R) \cdot t_X^* \cdot t^{\text{he}}_{\text{ns}}(L) \\ & - t^{\text{ee}}_{\text{sn}}(L) \cdot t_X \cdot r^{\text{eh}}_{\text{nn}}(R) \cdot t_X^* \cdot t^{\text{hh}}_{\text{ns}}(L) - t^{\text{eh}}_{\text{sn}}(L) \cdot t_X^* \cdot r^{\text{he}}_{\text{nn}}(R) \cdot t_X \cdot t^{\text{eh}}_{\text{ns}}(L) \Big], \end{aligned} \tag{4.81}$$

where t_X (t_X^*) is the transmission coefficient of a material X in the electron (hole) branch. Among the transport coefficients shown in Eqs. (4.39)–(4.46), it is possible to confirm the relations

$$t_{\rm ns}^{\rm ee}(L) \cdot \frac{\Delta}{\Omega} \cdot t_{\rm sn}^{\rm hh}(L) - t_{\rm ns}^{\rm eh}(L) \cdot \frac{\Delta}{\Omega} \cdot t_{\rm sn}^{\rm eh}(L) = 2r_{\rm nn}^{\rm eh}(L), \tag{4.82}$$

$$t_{\rm ns}^{\rm hh}(L) \cdot \frac{\Delta}{\Omega} \cdot t_{\rm sn}^{\rm ee}(L) - t_{\rm ns}^{\rm he}(L) \cdot \frac{\Delta}{\Omega} \cdot t_{\rm sn}^{\rm ee}(L) = 2r_{\rm nn}^{\rm he}(L). \tag{4.83}$$

These relationships hold true even after applying the analytic continuation $E \to i\hbar\omega_n$. Thus, we obtain

$$A_1 = \frac{2ie}{\hbar} \sum_p k_B T \sum_{\omega_n} \left[r_{\rm nn}^{\rm eh}(L) \cdot t_X^* \cdot r_{\rm nn}^{\rm he}(R) \cdot t_X - r_{\rm nn}^{\rm he}(L) \cdot t_X \cdot r_{\rm nn}^{\rm eh}(R) \cdot t_X^* \right]. \tag{4.84}$$

The reflection processes indicated by this equation are illustrated in Fig. 4.7b. The processes include the Andreev reflection once at the left interface and once at the right interface. The transport coefficients include t_n which is the transmission coefficients of the interface between the superconductor and the material X. To proceed with the calculation let us choose $t_n \ll 1$. The results

$$A_1 = \frac{e}{\hbar} \sum_p k_B T \sum_{\omega_n} |t_X t_n^2|^2 \left(\frac{\Delta}{\Omega_n} \right)^2 \sin\varphi \tag{4.85}$$

correspond to $J_1 \sin\varphi$. In the nth order term in Eq. (4.79) are derived from the processes including the Andreev reflection n times at the left interface and n times at the right interface. In addition, nth order process is proportional to $|t_X|^{2n}$ because a quasiparticle travel the X segment $2n$ times. When $|t_X|^2 \ll 1$ as it is in the insulator, the higher harmonics are negligible. In Eq. (4.78), the higher order terms contribute to the Josephson current because the two interfaces and the material X are highly transparent. Figure 4.7c shows the reflection process which contribute to $J_2 \sin(2\varphi)$.

Note

As already mentioned in the formula of Eq. (4.62) that the Josephson current is described by the Andreev reflection coefficients. Although the current should be gauge invariant, the description of the transport coefficients depends on a gauge choice. The user of Eq. (4.62) should pay attention to the gauge choice with which the Andreev reflection coefficients are calculated. In this text, Eq. (4.62) is correct when we calculate the transport coefficients in Eq. (4.69) from the wave functions in Eqs. (4.66) and (4.68).

References

Ambegaokar, V., Baratoff, A.: Phys. Rev. Lett. **10**, 486 (1963). https://doi.org/10.1103/PhysRevLett.10.486

Andreev, A.F.: Sov. Phys. JETP **19**, 1228 (1964)

Asano, Y.: Phys. Rev. B **61**, 1732 (2000). https://doi.org/10.1103/PhysRevB.61.1732

Blonder, G.E., Tinkham, M., Klapwijk, T.M.: Phys. Rev. B **25**, 4515 (1982). https://doi.org/10.1103/PhysRevB.25.4515

Furusaki, A., Tsukada, M.: Solid State Commun. **78**(4), 299 (1991). https://doi.org/10.1016/0038-1098(91)90201-6. http://www.sciencedirect.com/science/article/pii/0038109891902016

Kulik, I.O., Omel'yanchuk, A.N.: Sov. J. Low. Temp. Phys. **3**, 945 (1977)

Takane, Y., Ebisawa, H.: J. Phys. Soc. Jpn. **61**(5), 1685 (1992). https://doi.org/10.1143/JPSJ.61.1685

Chapter 5
Unconventional Superconductor

Abstract We will discuss the Andreev reflection from a superconductor that is characterized by an unconventional pair potential belonging to spin-singlet d-wave symmetry or spin-triplet p-wave symmetry. The unconventional pair potential changes its sign on the Fermi surface, which is a source of the Andreev bound states localizing at a surface of a superconductor. The resonant transmission of a Cooper pair via the Andreev bound states causes the zero-bias anomaly in the conductance spectra and the low-temperature anomaly in the Josephson current. Recently, a surface Andreev bound state is called topologically protected bound state at a surface of a topologically nontrivial superconductor. We will also explain how to topologically characterize unconventional superconductors briefly.

5.1 Pair Potentials in Unconventional Symmetry

In previous chapters, we have assumed that a superconductor is a simple metal such as Al, Pb, and Nb. A Cooper pair in such a conventional superconductor belongs to the spin-singlet s-wave symmetry class. In this chapter, we focus on the Andreev reflection by a superconductor belonging to unconventional symmetry class such as spin-singlet d-wave and spin-triplet p-wave. To describe unconventional superconducting state, we first generalize the pair potential within the mean-field theory,

$$\Delta_{\alpha,\beta}(\boldsymbol{r}_1, \boldsymbol{r}_2) = g(\boldsymbol{r}_1 - \boldsymbol{r}_2)\left\langle \psi_\alpha(\boldsymbol{r}_1)\psi_\beta(\boldsymbol{r}_2)\right\rangle = -g(\boldsymbol{r}_1 - \boldsymbol{r}_2)\left\langle \psi_\beta(\boldsymbol{r}_2)\psi_\alpha(\boldsymbol{r}_1)\right\rangle, \tag{5.1}$$

$$= -\Delta_{\beta,\alpha,}(\boldsymbol{r}_2, \boldsymbol{r}_1), \tag{5.2}$$

where the attractive interaction $g(\boldsymbol{r}_1 - \boldsymbol{r}_2) = g(\boldsymbol{r}_2 - \boldsymbol{r}_1)$ is symmetric under the permutation of real space coordinate and we have used the anticommutation relation of fermion operators. The pair potential must be antisymmetric under the permutation of two electrons as shown in Eq. (5.2) as a result of the Fermi-Dirac statistics of electrons. The negative sign on the right-hand side of Eq. (5.2) should be derived from the permutation of spins or the permutation of spatial coordinates of two electrons. This relation enables us to categorize superconductors into two classes: spin-singlet

Y. Asano, *Andreev Reflection in Superconducting Junctions*,
SpringerBriefs in Physics, https://doi.org/10.1007/978-981-16-4165-7_5

even-parity and spin-triplet odd-parity. By applying the Fourier transformation

$$\psi_\alpha(\boldsymbol{r}) = \frac{1}{\sqrt{V_{\rm vol}}}\sum_{\boldsymbol{k}} c_{\boldsymbol{k},\alpha} e^{i\boldsymbol{k}\cdot\boldsymbol{r}}, \quad g(\boldsymbol{r}) = \frac{1}{V_{\rm vol}}\sum_{\boldsymbol{k}} g_{\boldsymbol{k}} e^{i\boldsymbol{k}\cdot\boldsymbol{r}}, \tag{5.3}$$

the pair potential is represented as

$$\Delta_{\alpha,\beta}(\boldsymbol{R},\boldsymbol{\rho}) = \frac{1}{V_{\rm vol}^2}\sum_{\boldsymbol{q},\boldsymbol{k},\boldsymbol{k}'} g_{\boldsymbol{q}}\, e^{i\boldsymbol{q}\cdot\boldsymbol{\rho}}\, \langle c_{\boldsymbol{k},\alpha} c_{\boldsymbol{k}',\beta}\rangle e^{i(\boldsymbol{k}+\boldsymbol{k}')\cdot\boldsymbol{R}} e^{i(\boldsymbol{k}-\boldsymbol{k}')\cdot\boldsymbol{\rho}/2}, \tag{5.4}$$

$$\boldsymbol{R} = \frac{\boldsymbol{r}_1+\boldsymbol{r}_2}{2}, \quad \boldsymbol{\rho} = \boldsymbol{r}_1 - \boldsymbol{r}_2. \tag{5.5}$$

Here, we put $\boldsymbol{k}' = -\boldsymbol{k}$ because we consider a uniform superconductor. Namely, the pair potential is independent of the center-of-mass coordinate of two electrons $\boldsymbol{R}$. As a result, we have an expression of the pair potential,

$$\Delta_{\alpha,\beta}(\boldsymbol{\rho}) = \frac{1}{V_{\rm vol}^2}\sum_{\boldsymbol{k},\boldsymbol{q}} g_{\boldsymbol{q}} \langle c_{\boldsymbol{k},\alpha} c_{-\boldsymbol{k},\beta}\rangle e^{i(\boldsymbol{k}+\boldsymbol{q})\cdot\boldsymbol{\rho}} = \frac{1}{V_{\rm vol}}\sum_{\boldsymbol{k}} \Delta_{\alpha,\beta}(\boldsymbol{k}) e^{i\boldsymbol{k}\cdot\boldsymbol{\rho}}, \tag{5.6}$$

with the Fourier component

$$\Delta_{\alpha,\beta}(\boldsymbol{k}) = \frac{1}{V_{\rm vol}}\sum_{\boldsymbol{p}} g(\boldsymbol{k}-\boldsymbol{p}) f_{\alpha,\beta}(\boldsymbol{p}), \quad f_{\alpha,\beta}(\boldsymbol{k}) = \langle c_{\boldsymbol{k},\alpha} c_{-\boldsymbol{k},\beta}\rangle. \tag{5.7}$$

In the weak coupling theory, two electrons on the Fermi level form a Cooper pair due to an attractive interaction. Therefore, $\boldsymbol{k}$ and $\boldsymbol{p}$ in Eq. (5.7) are limited to be momenta on the Fermi surface. To proceed the argument, let us consider a superconductor in two dimensions and replace the summation by the integration,

$$\frac{1}{V_{\rm vol}}\sum_{\boldsymbol{k}} \to \int d\xi\, N(\xi) \int_0^{2\pi} \frac{d\theta}{2\pi}, \tag{5.8}$$

where $N(\xi)$ is the density of states per spin per volume and θ is an angle on the two-dimensional Fermi surface as shown in Fig. 5.1. The pair potential is described as

$$\Delta_{\alpha,\beta}(\theta) = \int d\xi\, N(\xi) \int_0^{2\pi} \frac{d\theta'}{2\pi}\, g(\theta,\theta')\, f_{\alpha,\beta}(\xi,\theta'). \tag{5.9}$$

The pairing function is decomposed into the Fourier series as

$$f_{\alpha,\beta}(\xi,\theta) = f_{\alpha,\beta}(\xi) + \sum_{n=1} f^{(c)}_{\alpha,\beta}(\xi,n)\cos(n\theta) + f^{(s)}_{\alpha,\beta}(\xi,n)\sin(n\theta). \tag{5.10}$$

The transformation of $\boldsymbol{k} \to -\boldsymbol{k}$ is described by $\theta \to \theta + \pi$. The component $f_{\alpha,\beta}(\xi)$ is the even-parity s-wave because it is independent of θ. The functions $\cos(2m\theta)$ and $\sin(2m\theta)$ are even-parity for m being an integer number, whereas $\cos((2m-1)\theta)$ and $\sin((2m-1)\theta)$ are odd-parity. When $g(\theta, \theta') = g$ is independent of θ, the pair potential is also independent of θ because the integral over θ' extracts the s-wave component from the pairing function. The resulting relation

$$\Delta_{\alpha,\beta} = \int d\xi N(\xi) g \, f_{\alpha,\beta}(\xi) \tag{5.11}$$

recovers the gap equation in the BCS theory. At $n = 1$, $f^{(c)}_{\alpha,\beta}(\xi, 1)\cos(\theta)$ and $f^{(s)}_{\alpha,\beta}(\xi, 1)\sin(\theta)$ belong, respectively, to p_x-symmetry and p_y-symmetry because of $k_x = k_F \cos\theta$ and $k_y = k_F \sin\theta$. To describe p-wave pair potential, we should choose

$$g(\theta, \theta') = 2\cos(\theta - \theta') = 2\cos\theta\cos\theta' + 2\sin\theta\sin\theta', \tag{5.12}$$

where the first (second) term extracts p_x-wave (p_y-wave) component from the pairing function. In the similar manner, the d-wave pair potential is described by choosing

$$g(\theta, \theta') = 2\cos(2\theta - 2\theta') = 2\cos(2\theta)\cos(2\theta)' + 2\sin(2\theta)\sin(2\theta'). \tag{5.13}$$

The first (second) term extracts $d_{x^2-y^2}$-wave (d_{xy}-wave) component from the pairing function. The unconventional pair potentials on the Fermi surface are illustrated in Fig. 5.1. The pair potential for an unconventional superconductor changes its sign on the Fermi surface, which enriches the transport phenomena as we will discuss in this chapter.

The BdG Hamiltonian in momentum space is given by

$$H_{\text{BdG}}(\boldsymbol{k}) = \begin{bmatrix} \xi_{\boldsymbol{k}}\,\hat{\sigma}_0 & \hat{\Delta}_{\boldsymbol{k}}\,e^{i\varphi} \\ -\hat{\Delta}^*_{-\boldsymbol{k}}\,e^{-i\varphi} & -\xi^*_{-\boldsymbol{k}}\,\hat{\sigma}_0 \end{bmatrix}. \tag{5.14}$$

The spin-singlet pair potential is decomposed as

$$\hat{\Delta}_{\boldsymbol{k}} = \Delta_{\boldsymbol{k}}\, i\hat{\sigma}_2, \quad \Delta_{\boldsymbol{k}} = \begin{cases} \Delta & : s\text{-wave} \\ \Delta(k_x^2 - k_y^2)/k_F^2 & : d_{x^2-y^2}\text{-wave} \\ \Delta(2k_x k_y)/k_F^2 & : d_{xy}\text{-wave} \end{cases}. \tag{5.15}$$

The spin-triplet pair potential can be described by

$$\hat{\Delta}_{\boldsymbol{k}} = \Delta_{\boldsymbol{k}}\, i\boldsymbol{d}\cdot\hat{\boldsymbol{\sigma}}\,\hat{\sigma}_2, \quad \Delta_{\boldsymbol{k}} = \begin{cases} \Delta k_x/k_F : p_x\text{-wave} \\ \Delta k_y/k_F : p_y\text{-wave} \end{cases}, \tag{5.16}$$

where $\boldsymbol{d}$ is a real unit vector in spin space representing a spin structure of a Cooper pair as

$$i\boldsymbol{d}\cdot\hat{\boldsymbol{\sigma}}\hat{\sigma}_2 = \begin{bmatrix} d_{\uparrow\uparrow} & d_{\uparrow\downarrow} \\ d_{\downarrow\uparrow} & d_{\downarrow\downarrow} \end{bmatrix} = \begin{bmatrix} id_2 - d_1 & d_3 \\ d_3 & id_2 + d_1 \end{bmatrix}. \tag{5.17}$$

In this book, $\Delta_{\boldsymbol{k}}$ represents the $\boldsymbol{k}$-dependence of the pair potential. For a spin-triplet superconductor, $\boldsymbol{d}$ represents the spin structure of a Cooper pair. We focus only on unitary states $\boldsymbol{d}\times\boldsymbol{d}^* = 0$ for spin-triplet pair potential. In such case, the BdG Hamiltonian is represented as

$$H_{\mathrm{BdG}}(\boldsymbol{k}) = \begin{bmatrix} \xi_{\boldsymbol{k}}\,\hat{\sigma}_0 & \Delta_{\boldsymbol{k}}\,\hat{\sigma}_S\,e^{i\varphi} \\ \Delta_{\boldsymbol{k}}\,\hat{\sigma}_S^{\dagger}\,e^{-i\varphi} & -\xi_{\boldsymbol{k}}\,\hat{\sigma}_0 \end{bmatrix}, \tag{5.18}$$

where we introduced a matrix in spin space,

$$\hat{\sigma}_S = \begin{cases} i\hat{\sigma}_2 & : \text{singlet} \\ i\boldsymbol{d}\cdot\hat{\boldsymbol{\sigma}}\hat{\sigma}_2 & : \text{triplet} \end{cases}, \tag{5.19}$$

and we used $\xi^*_{-\boldsymbol{k}} = \xi_{\boldsymbol{k}}$. The BdG equation can be solved as

$$H_{\mathrm{BdG}}(\boldsymbol{k})\,\check{\Phi} \begin{bmatrix} u_{\boldsymbol{k}}\,\hat{\sigma}_0 \\ v_{\boldsymbol{k}}\,s_{\boldsymbol{k}}\,\hat{\sigma}_S^{\dagger} \end{bmatrix} = \check{\Phi} \begin{bmatrix} u_{\boldsymbol{k}}\,\hat{\sigma}_0 \\ v_{\boldsymbol{k}}\,s_{\boldsymbol{k}}\,\hat{\sigma}_S^{\dagger} \end{bmatrix} E_{\boldsymbol{k}}, \tag{5.20}$$

$$E_{\boldsymbol{k}} = \sqrt{\xi_{\boldsymbol{k}}^2 + \Delta_{\boldsymbol{k}}^2}, \quad s_{\boldsymbol{k}} \equiv \frac{\Delta_{\boldsymbol{k}}}{|\Delta_{\boldsymbol{k}}|} \tag{5.21}$$

$$\check{\Phi} = \begin{bmatrix} e^{i\varphi/2}\hat{\sigma}_0 & 0 \\ 0 & e^{-i\varphi/2}\hat{\sigma}_0 \end{bmatrix}, \quad u_{\boldsymbol{k}} = \sqrt{\frac{1}{2}\left(1 + \frac{\xi_{\boldsymbol{k}}}{E_{\boldsymbol{k}}}\right)}, \quad v_{\boldsymbol{k}} = \sqrt{\frac{1}{2}\left(1 - \frac{\xi_{\boldsymbol{k}}}{E_{\boldsymbol{k}}}\right)}. \tag{5.22}$$

The sign of the pair potential $s_{\boldsymbol{k}}$ enters the wave function because the pair potential depends on the direction of the wavenumber on the Fermi surface. Here, we note several facts which we used on the way to the derivation. Pauli's matrices obey the relations

$$\hat{\boldsymbol{\sigma}}^* = -\hat{\sigma}_2\,\hat{\boldsymbol{\sigma}}\,\hat{\sigma}_2, \quad \boldsymbol{a}\cdot\hat{\boldsymbol{\sigma}}\;\boldsymbol{b}\cdot\hat{\boldsymbol{\sigma}} = \boldsymbol{a}\cdot\boldsymbol{b}\,\hat{\sigma}_0 + i\,\boldsymbol{a}\times\boldsymbol{b}\cdot\hat{\boldsymbol{\sigma}}. \tag{5.23}$$

For the orbital part of the pair potential, we have used the relations

$$\Delta^*_{-\boldsymbol{k}} = \begin{cases} \Delta_{\boldsymbol{k}} & : \text{singlet} \\ -\Delta_{\boldsymbol{k}} & : \text{triplet} \end{cases}, \tag{5.24}$$

which is true for the pair potentials in Fig. 5.1. However, the relation is not true for a chiral superconductor. The solution in Eq. (5.20) can be represented in different

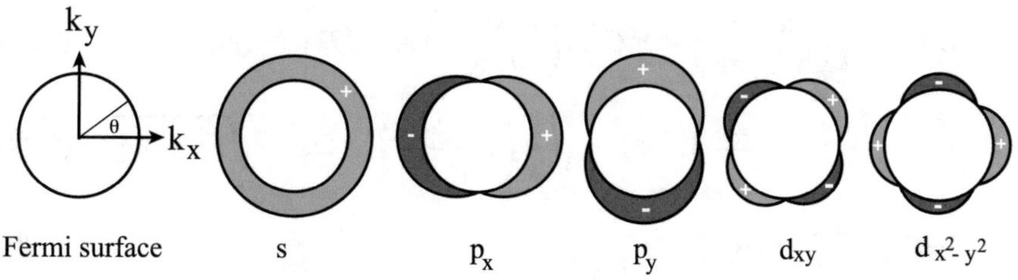

Fig. 5.1 The pair potentials are illustrated on the two-dimensional Fermi surface. An unfilled circle represents the Fermi sphere in two dimensions

ways. For example,

$$\check{\Phi} \begin{bmatrix} u_{\boldsymbol{k}}\, s_{\boldsymbol{k}}\, \hat{\sigma}_S \\ v_{\boldsymbol{k}}\, \hat{\sigma}_0 \end{bmatrix} \tag{5.25}$$

is also a solution of the BdG equation. The wave function depends on gauge choices for phase and choices of matrix structures in spin space. We will use Eq. (5.25) as a wave function at the hole branch in following sections.

5.2 Conductance in an NS Junction

We consider the conductance of an NS junction consisting of an unconventional superconductor in two dimensions. To calculate the reflection coefficients, we first describe the wave function in a normal metal,

$$\phi_{\mathrm{L}}(\boldsymbol{r}) = \left[\begin{pmatrix} \alpha \\ 0 \end{pmatrix} e^{ik_x x} + \begin{pmatrix} 0 \\ \beta \end{pmatrix} e^{-ik_x x} + \begin{pmatrix} A \\ 0 \end{pmatrix} e^{-ik_x x} + \begin{pmatrix} 0 \\ B \end{pmatrix} e^{ik_x x} \right] f_{k_y}(y), \tag{5.26}$$

where $f_{k_y}(y)$ is the wave function in the y-direction at the wavenumber being k_y and $k_x = \sqrt{k_F^2 - k_y^2}$ is the wavenumber in the x-direction on the Fermi surface. We have neglected the corrections to the wave number of the order of $\Delta/\epsilon_F \ll 1$. The amplitude of wave function,

$$X = \begin{bmatrix} X_\uparrow \\ X_\downarrow \end{bmatrix}, \tag{5.27}$$

has two spin components for X being α, β, A, and B in Eq. (5.26). This is also true for C and D in Eq. (5.28). Thus "0" in Eq. (5.26) is 2×1 null vector in spin space. The right-going wave in a superconductor is represented as

$$\phi_{\mathrm{R}}(\boldsymbol{r}) = \Phi\left[\begin{pmatrix} u_+\hat{\sigma}_0 \\ v_+ s_+ \hat{\sigma}_S^\dagger \end{pmatrix} e^{ik_x x} C + \begin{pmatrix} v_- s_- \hat{\sigma}_S \\ u_- \hat{\sigma}_0 \end{pmatrix} e^{-ik_x x} D\right] f_{k_y}(y), \tag{5.28}$$

$$u_\pm = \sqrt{\frac{1}{2}\left(1 + \frac{\Omega_\pm}{E}\right)}, \quad v_\pm = \sqrt{\frac{1}{2}\left(1 - \frac{\Omega_\pm}{E}\right)}, \quad \Omega_\pm = \sqrt{E^2 - \Delta_\pm^2}, \tag{5.29}$$

$$\Delta_\pm = \Delta(\pm k_x, k_y), \quad s_\pm = \frac{\Delta_\pm}{|\Delta_\pm|}, \tag{5.30}$$

where C and D represent the amplitudes of outgoing waves. In the electron branch, the wavenumber of a right-going wave is (k_x, k_y) on the Fermi surface. The wave function of an electron is a function of $\Delta_+ = \Delta(k_x, k_y)$. In the hole branch, on the other hand, the pair potential is $\Delta_- = \Delta(-k_x, k_y)$ because the wavenumber on the Fermi surface is $(-k_x, k_y)$. As we see below, the sign of the pair potential $s_\pm$ plays an important role in quantum transport phenomena.

Using the boundary conditions in Eqs. (4.31) and (4.32), it is possible to derive a relation

$$\left[\bar{k}\tau_3 U^{-1} - U^{-1}\bar{k}\tau_3 + 2iz_0 U^{-1}\right]\begin{bmatrix}\alpha \\ \beta\end{bmatrix} = -\left[\bar{k}\tau_3 U^{-1} + U^{-1}\bar{k}\tau_3 + 2iz_0 U^{-1}\right]\begin{bmatrix}A \\ B\end{bmatrix},$$

$$U = \Phi\begin{bmatrix} u_+\hat{\sigma}_0 & v_- s_- \hat{\sigma}_S \\ v_+ s_+ \hat{\sigma}_S^\dagger & u_- \hat{\sigma}_0 \end{bmatrix}. \tag{5.31}$$

The reflection coefficients defined by

$$\begin{bmatrix}A \\ B\end{bmatrix} = \begin{bmatrix}\hat{r}_{\mathrm{ee}} & \hat{r}_{\mathrm{eh}} \\ \hat{r}_{\mathrm{he}} & \hat{r}_{\mathrm{hh}}\end{bmatrix}\begin{bmatrix}\alpha \\ \beta\end{bmatrix} \tag{5.32}$$

are calculated as

$$\hat{r}_{\mathrm{ee}} = \frac{r_n(1 - \Gamma_+\Gamma_-)}{1 - |r_n|^2\Gamma_+\Gamma_-}\hat{\sigma}_0, \quad \hat{r}_{\mathrm{hh}} = \frac{r_n^*(1 - \Gamma_+\Gamma_-)}{1 - |r_n|^2\Gamma_+\Gamma_-}\hat{\sigma}_0, \tag{5.33}$$

$$\hat{r}_{\mathrm{he}} = \frac{|t_n|^2\Gamma_+ e^{-i\varphi}}{1 - |r_n|^2\Gamma_+\Gamma_-}\hat{\sigma}_S^\dagger, \quad \hat{r}_{\mathrm{eh}} = \frac{|t_n|^2\Gamma_- e^{i\varphi}}{1 - |r_n|^2\Gamma_+\Gamma_-}\hat{\sigma}_S, \tag{5.34}$$

with the transmission coefficients in the normal state in Eq. (4.37) and

$$\Gamma_\pm \equiv \frac{v_\pm}{u_\pm} s_\pm = \frac{\Delta_\pm}{E + \Omega_\pm}. \tag{5.35}$$

In an unconventional superconductor junction, the two pair potentials Δ_+ and Δ_- enter the transport coefficients. In particular, the Andreev reflection at $E = 0$ shows qualitatively different behaviors depending on the relative sign between the two pair potentials. To see this, let us classify the pair potentials in Fig. 5.1 into two groups.

The pair potentials can be described as

$$\Delta_+ = s_+|\Delta_{\boldsymbol{k}}|, \quad \Delta_- = s_-|\Delta_{\boldsymbol{k}}|. \tag{5.36}$$

We find that

$$\begin{aligned} s_+ s_- &= 1 \quad : s,\ p_y,\ \text{and } d_{x^2-y^2}, \\ s_+ s_- &= -1 : d_{xy} \text{ and } p_x \end{aligned} \tag{5.37}$$

are satisfied for all the propagating channels k_y. It is possible to obtain simple expression of the Andreev reflection coefficients in these pairing symmetries because $\Omega_\pm = \Omega$, $u_\pm = u$, and $v_\pm = v$ are satisfied. In the limit of $E \to 0$, we note that the function

$$\lim_{E\to 0} \Gamma_\pm = -is_\pm \tag{5.38}$$

is proportional to the sign of the pair potential.

In the case of $s_+s_- = 1$, the reflection probabilities at $E = 0$ in the tunneling limit $|t_n| \ll 1$ become

$$\lim_{E\to 0} |\hat{r}_{\text{he}}|^2 = \left(\frac{|t_n|^2}{2}\right)^2 \hat{\sigma}_0, \quad \lim_{E\to 0} |\hat{r}_{\text{ee}}|^2 = |r_n|^2 \hat{\sigma}_0. \tag{5.39}$$

The zero-bias conductance is proportional to $|t_n|^4$ because the conductance is calculated by substituting the reflection coefficients into the formula in Eq. (4.52). The conductance for $s_+s_- = 1$ can be represented by Eqs. (4.53) and (4.54) with $\Delta \to \Delta_+$. The zero-bias conductance is zero in the tunnel limit $|t_n| \ll 1$. In Fig. 5.2b and d, we show the differential conductance of NS junctions consisting of a $d_{x^2-y^2}$-wave superconductor and that of a p_y-wave superconductor, respectively. The conductance spectra for $z_0 = 5$ correspond to the density of states at the uniform superconducting state in both Fig. 5.2b and d. The V-shaped spectra around zero bias are a result of nodes of the superconducting gap on the Fermi surface.

In the case of $s_+s_- = -1$, on the other hand, we find $\Gamma_+\Gamma_- = 1$ at $E = 0$, which results in

$$\lim_{E\to 0} |\hat{r}_{\text{he}}|^2 = \hat{\sigma}_0, \quad \lim_{E\to 0} |\hat{r}_{\text{ee}}|^2 = 0. \tag{5.40}$$

The Andreev reflection is perfect at $E = 0$ independent of the potential barrier at the interface. The differential conductance for $s_+s_- = -1$ is calculated to be

$$G_{\text{NS}} = \frac{2e^2}{h} \sum_{k_y} \left. \frac{2|t_n|^4 \Delta_+^2}{4(1-|t_n|^2)E^2 + |t_n|^4 \Delta_+^2} \right|_{E=eV} \quad \text{for } 0 \le eV < |\Delta_+|. \tag{5.41}$$

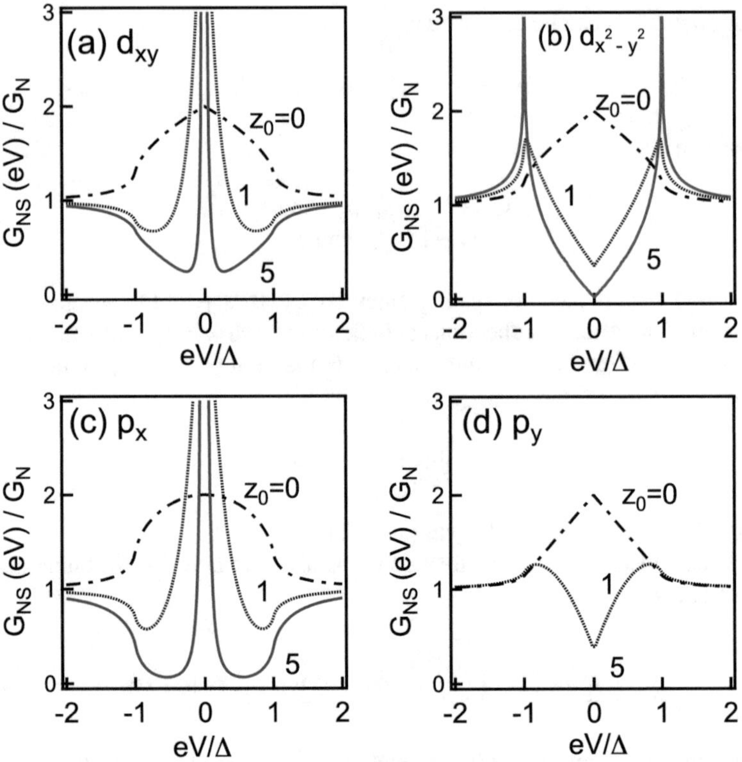

Fig. 5.2 The conductance spectra are calculated for an NS junction consisting of an unconventional superconductor for several choices of the barrier parameter z_0. The transmission probability of a junction in the normal state is unity at $z_0 = 0$. The results for $z_0 = 5$ correspond to the tunnel spectra measured in STM/STS experiments, where the transmission probability of the potential barrier is estimated to be 0.026

At $E = eV = 0$, the conductance is twice of the Sharvin conductance. In the limit of $|t_n|^2 \ll 1$ and $E \ll |\Delta_+|$, the expression

$$G_{\rm NS} \approx \frac{4e^2}{h} \sum_{k_y} \left. \frac{(|t_n|^2 \Delta_+/2)^2}{E^2 + (|t_n|^2 \Delta_+/2)^2} \right|_{E=eV} \tag{5.42}$$

suggests that conductance has a peak at zero bias and the peak width is given by $|t_n|^2|\Delta_+|/2$. For $eV \geq |\Delta_+|$, the conductance for $s_+ s_- = -1$ is calculated to be

$$G_{\rm NS} = \frac{2e^2}{h} \sum_{k_y} \left. \frac{2|t_n|^2 \left[|t_n|^2 E^2 + (2 - |t_n|^2)\sqrt{E^2 - \Delta_+^2}\, E \right]}{\left[(2 - |t_n|^2)E + |t_n|^2 \sqrt{E^2 - \Delta_+^2} \right]^2} \right|_{E=eV} . \tag{5.43}$$

In Fig. 5.2a and c, we show the differential conductance of a d_{xy}-wave junction and that of a p_x-wave junction, respectively. The conductance in the tunnel limit shows a large peak at zero bias for both (a) and (c). (Tanaka 1995) The sign change of pair potential is a general feature of all unconventional superconductors. Thus, the zero-bias anomaly observed in STM/STS experiments in a superconductor suggests that the pairing symmetry of the superconductor would be unconventional.

5.3 Surface Andreev Bound States

The reasons of the zero-bias anomaly appearing in the conductance spectra should be clarified (Asano et al. 2004). In Fig. 5.3, the Andreev reflection process is decomposed into a series of multiple reflections by the pair potential and those by the barrier potential. In the electron branch, the transmission and reflection coefficients of the barrier in the normal state are t_n and r_n, respectively. In the hole branch, they are given by t_n^* and r_n^*. At the first-order process, an electron transmits into the superconductor with the coefficient t_n. For $0 \leq E < \Delta$, the penetration of such an electron into a superconductor is limited spatially by the coherence length ξ_0 from the interface. Thus, an electron is reflected as a hole by the pair potential Δ_+. The Andreev reflection coefficients in this case is obtained by putting $t_n = 1$ into Eq. (5.34) as

$$\hat{r}_{\rm he}^{(0)} = \Gamma_+ e^{-i\varphi} \hat{\sigma}_S^\dagger, \quad \hat{r}_{\rm eh}^{(0)} = \Gamma_- e^{i\varphi} \hat{\sigma}_S. \tag{5.44}$$

Finally, a reflected hole transmits back into the normal metal with t_n^*. The Andreev reflection coefficient at the first-order process is summarized as

$$\hat{r}_{\rm he}^{(1)} = t_n^* \cdot \hat{r}_{\rm he}^{(0)} \cdot t_n. \tag{5.45}$$

The reflection sequence is ordered from the right to the left of this equation. In the second-order process, a quasiparticle in the first process is reflected by the pair potential twice more. Simultaneously, a quasiparticle is reflected by the barrier potential twice. The resulting reflection coefficient becomes

$$\hat{r}_{\rm he}^{(2)} = t_n^* \left[\hat{r}_{\rm he}^{(0)} \cdot r_n \cdot \hat{r}_{\rm eh}^{(0)} \cdot r_n^* \right] \hat{r}_{\rm he}^{(0)} \cdot t_n. \tag{5.46}$$

By repeating the similar argument, it is possible to calculate mth-order reflection coefficient as

$$\hat{r}_{\rm he}^{(m)} = t_n^* \cdot \left[\hat{r}_{\rm he}^{(0)} \cdot r_n \cdot \hat{r}_{\rm eh}^{(0)} \cdot r_n^* \right]^{m-1} \hat{r}_{\rm he}^{(0)} \cdot t_n. \tag{5.47}$$

Due to Eq. (5.38), the Andreev reflection probability at $E = 0$ is given by

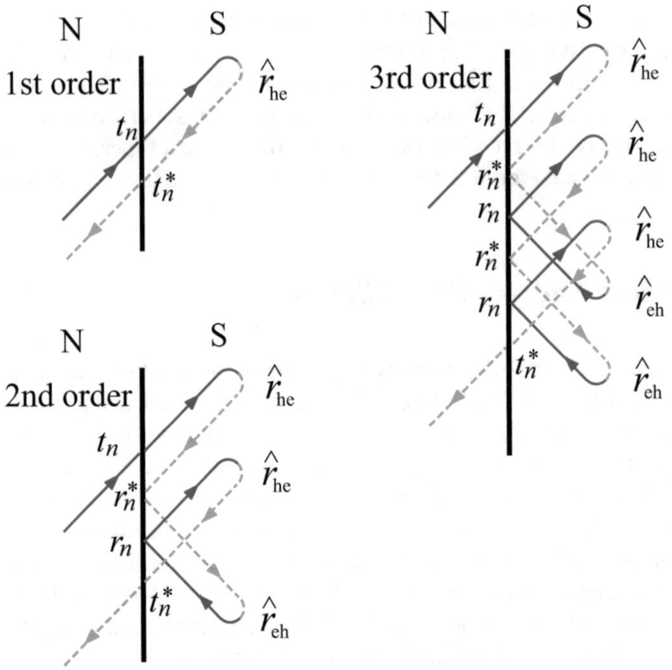

Fig. 5.3 The Andreev reflection process is decomposed into a series of reflections by the pair potential and by the barrier potential

$$\hat{r}_{\rm he}\,\hat{r}_{\rm he}^{\dagger} = |t_n|^4 \left| \sum_{m=0}^{\infty} (1-|t_n|^2)^m(-s_+s_-)^m \right|^2 \hat{\sigma}_0 = \frac{|t_n|^4\hat{\sigma}_0}{|1+(1-|t_n|^2)s_+s_-|^2}. \tag{5.48}$$

When the two pair potentials have the same sign (i.e., $s_+s_- = 1$), the Andreev reflection probability in Eq. (5.48) is represented by the summation of an alternating series. The zero-bias conductance represented as

$$G_{\rm NS}|_{E=0} = \frac{2e^2}{h} \sum_{k_y} \frac{2|t_n|^4}{(2-|t_n|^2)^2} \tag{5.49}$$

vanishes in the tunnel limit $|t_n| \to 0$. When the two pair potentials have the opposite sign (i.e., $s_+s_- = -1$), on the other hand, the Andreev reflection probability is unity independent of the potential barrier. The resulting conductance

$$G_{\rm NS}|_{E=0} = \frac{2e^2}{h} 2N_c \tag{5.50}$$

is the twice of the Sharvin conductance.

The relative sign of the two pair potentials $s_+ s_-$ plays a crucial role in the interference effect of a quasiparticle near a surface of a superconductor. Namely, $s_+ s_- = 1$ causes the destructive interference effect. The constructive interference effect with $s_+ s_- = -1$ enables the resonant transmission of a quasiparticle (resonant conversion of an electron to a Cooper pair) through the potential barrier. Thus, it is expected that the resonant states at $E = 0$ would be bound at the surface of an unconventional superconductor, (Buchholtz and Zwicknagl 1981, Hara and Nagai 1986). The zero-bias peak in the conductance could be interpreted as a result of the enhanced local density of states at the surface. It is possible to confirm the validity of the argument by seeking a bound state in terms of the wave function in Eq. (5.28) which represents the right-going wave in the superconductor at E,

$$\phi_{\mathrm{R}}(x) = \left[\begin{bmatrix} \hat{\sigma}_0 \\ \Gamma_+ \hat{\sigma}_S^\dagger \end{bmatrix} e^{ik_x x} C + \begin{bmatrix} \Gamma_- \hat{\sigma}_S \\ \hat{\sigma}_0 \end{bmatrix} e^{-ik_x x} D \right] e^{-x/\xi_0}. \tag{5.51}$$

Here, we assume that $E < |\Delta_\pm|$ and $\varphi = 0$. The imaginary part of wavenumber is represented explicitly to describe bound states at the surface. At $x = 0$, the wave function satisfies the boundary condition of

$$\phi_{\mathrm{R}}(0) = \begin{bmatrix} \hat{\sigma}_0 & \Gamma_- \hat{\sigma}_S \\ \Gamma_+ \hat{\sigma}_S^\dagger & \hat{\sigma}_0 \end{bmatrix} \begin{bmatrix} C \\ D \end{bmatrix} = 0. \tag{5.52}$$

The condition is satisfied when

$$\Gamma_+ \Gamma_- = 1. \tag{5.53}$$

In the case of $s_+ s_- = 1$, $E = \Delta_+$ is a solution of this equation. The wave function of such solution, however, is not localized at the interface. In the case of $s_+ s_- = -1$, $E = 0$ is a solution of this equation. The wave function of such a bound state is described by

$$\phi_{\mathrm{R}}(x) = C \sin(2k_x x)\, e^{-x/\xi_0} \begin{bmatrix} \hat{\sigma}_0 \\ -i\hat{\sigma}_S^\dagger s_+ \end{bmatrix}, \tag{5.54}$$

where C is a constant. Today, the surface Andreev bound states are called topologically protected bound states at a surface of a topologically nontrivial superconductor. We will address this issue in Sect. 5.5.

5.4 The Josephson Current in an SIS Junction

The Andreev bound states at a surface cause the zero-bias anomaly in the differential conductance in an NS junction. In this section, we discuss the effects of the surface bound states on the Josephson current. Let us consider an SIS junction, where two unconventional superconductors are identical to each other. In the left superconductor, the wave function is given by

$$\begin{aligned}\phi_{\mathrm{L}}(\boldsymbol{r}) =&\check{\Phi}_{\mathrm{L}}\left[\begin{pmatrix} u_{+}\hat{\sigma}_0 \\ v_{+}s_{+}\hat{\sigma}_S^{\dagger} \end{pmatrix} e^{ik_x x}\alpha + \begin{pmatrix} v_{-}s_{-}\hat{\sigma}_S \\ u_{-}\hat{\sigma}_0 \end{pmatrix} e^{-ik_x x}\beta\right.\\ &\left.+\begin{pmatrix} u_{-}\hat{\sigma}_0 \\ v_{-}s_{-}\hat{\sigma}_S^{\dagger} \end{pmatrix} e^{-ik_x x}A + \begin{pmatrix} v_{+}s_{+}\hat{\sigma}_S \\ u_{+}\hat{\sigma}_0 \end{pmatrix} e^{ik_x x}B\right] f_{k_y}(y),\end{aligned} \tag{5.55}$$

$$u_{\pm} = \sqrt{\frac{1}{2}\left(1+\frac{\Omega_{n,\pm}}{\hbar\omega_n}\right)}, \quad v_{\pm} = \sqrt{\frac{1}{2}\left(1-\frac{\Omega_{n,\pm}}{\hbar\omega_n}\right)}, \quad \Omega_{n,\pm} = \sqrt{\hbar^2\omega_n^2 + \Delta_{\pm}^2}, \tag{5.56}$$

with Eq. (5.30). In the same way, the wave function on a superconductor on the right is represented by

$$\phi_{\mathrm{R}}(\boldsymbol{r}) =\check{\Phi}_{\mathrm{R}}\left[\begin{pmatrix} u_{+}\hat{\sigma}_0 \\ v_{+}s_{+}\hat{\sigma}_S^{\dagger} \end{pmatrix} e^{ik_x x}C + \begin{pmatrix} v_{-}s_{-}\hat{\sigma}_S \\ u_{-}\hat{\sigma}_0 \end{pmatrix} e^{-ik_x x}D\right] f_{k_y}(y). \tag{5.57}$$

The phase of superconductor is represented by

$$\check{\Phi}_j = \begin{bmatrix} e^{i\varphi_j/2}\hat{\sigma}_0 & \hat{0} \\ \hat{0} & e^{-i\varphi_j/2}\hat{\sigma}_0 \end{bmatrix}, \tag{5.58}$$

with $j = L$ or R. The Josephson current can be calculated based on a formula (Tanaka and Kashiwaya 1996; Asano 2001)

$$J = \frac{e}{2\hbar}k_B T \sum_{\omega_n}\sum_{k_y} \mathrm{Tr}\left[\frac{\Delta_{+}\hat{\sigma}_S\hat{r}_{\mathrm{he}}}{\Omega_{n,+}} - \frac{\Delta_{-}\hat{\sigma}_S^{\dagger}\hat{r}_{\mathrm{eh}}}{\Omega_{n,-}}\right]. \tag{5.59}$$

As discussed at the end of Chap. 4, the expression of the formula depends on the gauge choice. The formula is correct when the Andreev reflection coefficients are calculated from the wave functions in Eqs. (5.55) and (5.57).

Applying the boundary conditions in Eqs. (4.31) and (4.32), we obtain the Andreev reflection coefficients. As shown in Eq. (5.37), we consider two cases to make clear the effects of the surface Andreev bound states at zero energy on the Josephson effect. Since $\Omega_{n\pm} = \Omega_n$, reflection coefficients result in

$$\hat{r}_{\text{he}} = \frac{\Delta_+ \hat{\sigma}_S^\dagger}{2i} \left[\frac{|t_n|^2 \{(\cos\varphi - 1)\,\hbar\omega_n + i\sin\varphi\,\Omega_n\} + |r_n|^2 (1 - s_+ s_-)\,\hbar\omega_n}{|t_n|^2 \left\{(\hbar\omega_n)^2 + \Delta_k^2 \cos^2(\varphi/2)\right\} + |r_n|^2 \left\{\hbar\omega_n \frac{1-s_+s_-}{2} + \Omega_n \frac{1+s_+s_-}{2}\right\}^2} \right],$$

$$\hat{r}_{\text{eh}} = \frac{\Delta_- \hat{\sigma}_S}{2i} \left[\frac{|t_n|^2 \{(\cos\varphi - 1)\,\hbar\omega_n - i\sin\varphi\,\Omega_n\} + |r_n|^2 (1 - s_+ s_-)\,\hbar\omega_n}{|t_n|^2 \left\{(\hbar\omega_n)^2 + \Delta_k^2 \cos^2(\varphi/2)\right\} + |r_n|^2 \left\{\hbar\omega_n \frac{1-s_+s_-}{2} + \Omega_n \frac{1+s_+s_-}{2}\right\}^2} \right],$$

with $\varphi = \varphi_L - \varphi_R$ and the transport coefficients in the normal state in Eq. (4.37). When $s_+ s_- = 1$ is satisfied in s-, $d_{x^2-y^2}$-, and p_y-wave junctions, we obtain

$$\hat{r}_{\text{he}} = \frac{\hat{\sigma}_S^\dagger |t_n|^2 \Delta_+}{2i} \left[\frac{(\cos\varphi - 1)\,\hbar\omega_n + i\sin\varphi\,\Omega_n}{\hbar^2\omega_n^2 + \Delta_k^2 \{1 - |t_n|^2 \sin^2(\varphi/2)\}} \right], \tag{5.60}$$

$$\hat{r}_{\text{eh}} = \frac{\hat{\sigma}_S |t_n|^2 \Delta_-}{2i} \left[\frac{(\cos\varphi - 1)\,\hbar\omega_n - i\sin\varphi\,\Omega_n}{\hbar^2\omega_n^2 + \Delta_k^2 \{1 - |t_n|^2 \sin^2(\varphi/2)\}} \right]. \tag{5.61}$$

The Josephson current becomes

$$J = \frac{e}{\hbar} \sin\varphi \sum_{k_y} \frac{|t_n|^2 \Delta_k}{2\sqrt{1 - |t_n|^2 \sin^2(\varphi/2)}} \tanh\left[\frac{\Delta_k \sqrt{1 - |t_n|^2 \sin^2(\varphi/2)}}{2k_B T} \right]. \tag{5.62}$$

The results are qualitatively same as the Josephson current in Eq. (4.76). For $s_+ s_- = -1$ in d_{xy}- and p_x-wave junctions, on the other hand, the coefficients are represented as

$$\hat{r}_{\text{he}} = \frac{\hat{\sigma}_S^\dagger \Delta_+}{2i} \left[\frac{|t_n|^2 \{(\cos\varphi - 1)\,\hbar\omega_n + i\sin\varphi\,\Omega_n\} + 2|r_n|^2 \hbar\omega_n}{\hbar^2\omega_n^2 + \Delta_k^2 |t_n|^2 \cos^2(\varphi/2)} \right], \tag{5.63}$$

$$\hat{r}_{\text{eh}} = \frac{\hat{\sigma}_S \Delta_-}{2i} \left[\frac{|t_n|^2 \{(\cos\varphi - 1)\,\hbar\omega_n - i\sin\varphi\,\Omega_n\} + 2|r_n|^2 \hbar\omega_n}{\hbar^2\omega_n^2 + \Delta_k^2 |t_n|^2 \cos^2(\varphi/2)} \right]. \tag{5.64}$$

The Josephson current is calculated to be

$$J = \frac{e}{\hbar} \sin\varphi \sum_{k_y} \frac{|t_n| \Delta_k}{2\cos(\varphi/2)} \tanh\left[\frac{\Delta_k |t_n| \cos(\varphi/2)}{2k_B T} \right]. \tag{5.65}$$

At $T = 0$, the current-phase relationship is fractional $J \propto \sin(\varphi/2)$ irrespective of t_n because of the resonant transmission of a Cooper pair via the Andreev bound states at the interface. In Fig. 5.4a, we plot the amplitude of the Josephson current as a function of temperature for p_x-, p_y-, $d_{x^2-y^2}$- and d_{xy}-wave SIS junctions. The results for p_x- and d_{xy}-wave junctions show that the Josephson current increases rapidly with the decrease of temperature for $T < 0.5T_c$. In Fig. 5.4b, we show the current-phase relationship in a p_x-wave SIS junction. The relation is sinusoidal at a

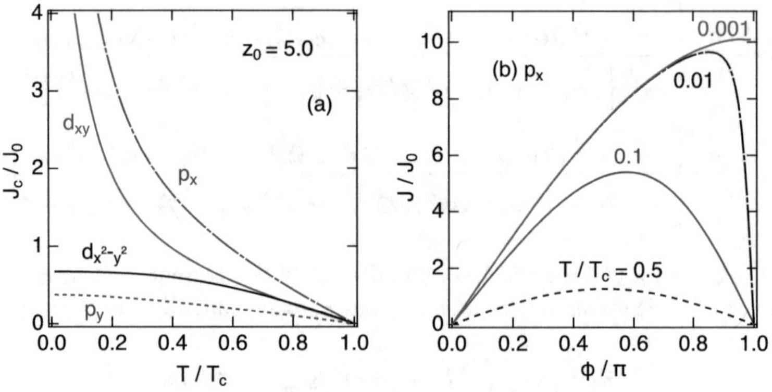

Fig. 5.4 The Josephson critical current is plotted as a function of temperature for p_x-, p_y-, $d_{x^2-y^2}$-, and d_{xy}-wave symmetry in (**a**). The current-phase relationship in a p_x-wave SIS junction is shown for several choices of temperatures in (**b**). The transmission probability of the barrier is about 0.026 as a result of choosing $z_0 = 5$

temperature near T_c, whereas it becomes factional at a low temperature $T = 0.001T_c$. Such unusual behavior is called low-temperature anomaly of the Josephson effect (Tanaka and Kashiwaya 1996).

It is possible to calculate the Josephson current from the energy spectra $E(\varphi)$ of bound states at the interface of an SIS junction based on the relation

$$J = \frac{2e}{\hbar}\frac{d}{d\varphi}E(\varphi). \tag{5.66}$$

Here, we discuss how the surface Andreev bound states at zero energy affects the bound state energy at the junction interface $E(\varphi)$. We focus on simple cases shown in Eq. (5.36) and consider 2×2 the BdG Hamiltonian by block diagonalizing the original 4×4 Hamiltonian into two spin sectors. In the left superconductor, the wave function localizing at the interface is described by

$$\phi_{\mathrm{L}}(\boldsymbol{r}) = \check{\Phi}_{\mathrm{L}}\left[\begin{pmatrix} E + i\Omega \\ \Delta_- \end{pmatrix} e^{-ik_x x} A + \begin{pmatrix} E - i\Omega \\ \Delta_+ \end{pmatrix} e^{ik_x x} B\right] e^{x/\xi_0} f_{k_y}(y), \tag{5.67}$$

with $\Omega = \sqrt{\Delta_{\boldsymbol{k}}^2 - E^2}$. The wave function on the right-hand side becomes

$$\phi_{\mathrm{R}}(\boldsymbol{r}) = \check{\Phi}_{\mathrm{R}}\left[\begin{pmatrix} E + i\Omega \\ \Delta_+ \end{pmatrix} e^{ik_x x} C + \begin{pmatrix} E - i\Omega \\ \Delta_- \end{pmatrix} e^{-ik_x x} D\right] e^{-x/\xi_0} f_{k_y}(y). \tag{5.68}$$

They are connected with each other at $x = 0$ by the boundary conditions in Eqs. (4.31) and (4.32). From the Boundary conditions, we obtain the equation

$$\begin{bmatrix} \chi\bar{k}(E+i\Omega) & -\chi\bar{k}(E-i\Omega) & (\bar{k}-2iz_0)(E+i\Omega) & -(\bar{k}+2iz_0)(E-i\Omega) \\ \chi^*\bar{k}\Delta_- & -\chi^*\bar{k}\Delta_+ & (\bar{k}-2iz_0)\Delta_+ & -(\bar{k}+2iz_0)\Delta_- \\ \chi(E+i\Omega) & \chi(E-i\Omega) & -(E+i\Omega) & -(E-i\Omega) \\ \chi^*\Delta_- & \chi^*\Delta_+ & -\Delta_+ & -\Delta_- \end{bmatrix} \begin{bmatrix} A \\ B \\ C \\ D \end{bmatrix} = 0, \tag{5.69}$$

with $\chi = e^{i\varphi/2}$. The determinant of the matrix is proportional to

$$|t_n|^2 \left[-|E+i\Omega|^2 \cos\varphi + E^2 - \Omega^2 \right] + |r_n|^2 \left[E^2 - \Omega^2 - s_+ s_- |E+i\Omega|^2 \right] = 0. \tag{5.70}$$

The energy of the bound states is calculated to be

$$E = \pm|\Delta_{\boldsymbol{k}}| \left[|t_n|^2 \cos^2\left(\frac{\varphi}{2}\right) + \frac{1+s_+s_-}{2}(1-|t_n|^2) \right]^{1/2}. \tag{5.71}$$

The negative sign must be chosen for occupied bound states which carry the Josephson current. In the case of s-wave junction (i.e., $s_+s_- = 1$), the energy of the Andreev bound states becomes

$$E = -\Delta \left[1 - |t_n|^2 \sin^2\left(\frac{\varphi}{2}\right) \right]^{1/2}. \tag{5.72}$$

The Josephson current calculated by using Eq. (5.66) is identical to Eq. (5.62) at $T = 0$. The energy of the Andreev bound states for a p_x-wave junction (i.e., $s_+s_- = -1$) results in

$$E = -|\Delta_{\boldsymbol{k}}||t_n| \cos\left(\frac{\varphi}{2}\right). \tag{5.73}$$

The Josephson current calculated from Eq. (5.66) is identical to Eq. (5.65) at $T = 0$. The energy spectra of the Andreev bound states at a junction interface are plotted as a function of the phase difference in Fig. 5.5. In both Eqs. (5.72) and (5.73), the occupied energy branch takes its minimum at $\varphi = 0$.

5.5 Topological Classification

A topologically nontrivial state of the material is a novel concept in condensed matter physics. Some of the time, the quantum states of an electron in a solid can be characterized in terms of topological numbers. When we apply the periodic boundary condition to obtain a solution of a Schroedinger equation, the wavenumber $\boldsymbol{k}$ indicates the place of the Brillouin zone and the wave function $\phi_l(\boldsymbol{k})$ describes properties of the quantum state labeled by l. In other words, $\phi_l(\boldsymbol{k})$ is a map from the Brillouin

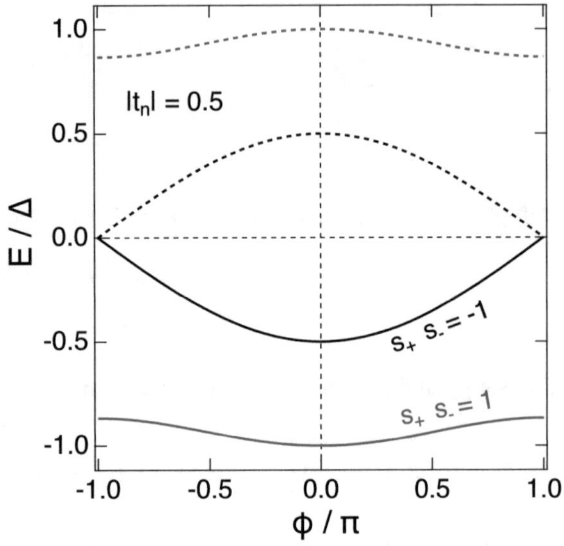

Fig. 5.5 The energy of Andreev bound states at a Josephson junction for $|t_n| = 0.5$. The results for $s_+s_- = 1$ in Eq. (5.72) are 2π periodic. On the other hand, the results for $s_+s_- = -1$ in Eq. (5.73) are 4π periodic. At $\varphi = \pi$, two bound states are degenerate at zero energy

zone to the Hilbert space. In this section, we will calculate a topological number (winding number) to characterize nodal unconventional superconductors preserving time-reversal symmetry. Before turning into details, important features of topological numbers are summarized,

1. To define a topological number, the electronic states must have gapped energy spectra. The Fermi level must lie in the gap.
2. Topological number $\mathcal{Z}$ is an integer number that can be defined in terms of the wave functions of all occupied states below the gap.
3. Topological number remains unchanged as long as the gap opens.

As the energy spectra are gapped at the Fermi level, topological materials are insulating or superconducting. These features enable us to understand why zero-energy states appear at a surface of a topologically nontrivial material as shown in Fig. 5.6a. The topological number is $\mathcal{Z} \neq 0$ (nontrivial) in the topological material at $x > 0$ and it is zero (trivial) in vacuum $x < 0$. Since $\mathcal{Z}$ is an integer number, it jumps at $x = 0$ discontinuously. To change the topological number, the gap must close at the surface. More specifically, electronic states at the Fermi level are necessary at $x = 0$. This argument explains the existence of metallic states (zero-energy states) at a surface of a topological material. In addition, the number of such surface bound states at the Fermi level is identical to $|\mathcal{Z}|$. This fact is called bulk-boundary correspondence.

In what follows, we explain how to topologically characterize nodal superconductors by focusing on a spin-triplet p_x-wave superconductor in two dimensions. The mean-field Hamiltonian is given by

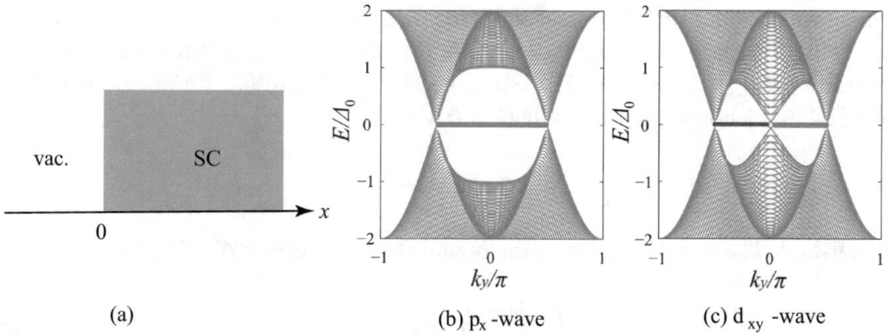

Fig. 5.6 A topologically nontrivial superconductor has a surface at $x = 0$ in (**a**). The energy eigenvalues of a p_x-wave superconductor and those of a d_{xy}-wave superconductor are plotted as a function of k_y in (**b**) and (**c**), respectively. The hard-wall boundary condition is applied in the x-direction and the periodic boundary condition is applied in the y-direction. As a result of nontrivial winding numbers in a superconductor, zero-energy states appear between the nodes of the pair potential

$$\mathcal{H} = \int d\boldsymbol{r}\ [\psi^\dagger(\boldsymbol{r}), \psi(\boldsymbol{r})] \begin{bmatrix} \xi_(\boldsymbol{r}) & \Delta_0 \frac{-\partial_x}{ik_F} \\ \Delta_0 \frac{-\partial_x}{ik_F} & -\xi^*(\boldsymbol{r}) \end{bmatrix} \begin{bmatrix} \psi(\boldsymbol{r}) \\ \psi^\dagger(\boldsymbol{r}) \end{bmatrix}, \tag{5.74}$$

and the BdG equation becomes

$$\begin{bmatrix} \xi_(\boldsymbol{r}) & \Delta_0 \frac{-\partial_x}{ik_F} \\ \Delta_0 \frac{-\partial_x}{ik_F} & -\xi^*(\boldsymbol{r}) \end{bmatrix} \begin{bmatrix} u_\nu(\boldsymbol{r}) \\ v_\nu(\boldsymbol{r}) \end{bmatrix} = E_\nu \begin{bmatrix} u_\nu(\boldsymbol{r}) \\ v_\nu(\boldsymbol{r}) \end{bmatrix}, \tag{5.75}$$

where the superconducting phase is fixed at $\varphi = 0$. We consider 2×2 BdG equation by focusing on spin $\uparrow$ sector with choosing $\hat{\sigma}_S = \hat{\sigma}_3$ in Eq. (5.19). The Bogoliubov transformation becomes

$$\begin{bmatrix} \psi(\boldsymbol{r}) \\ \psi^\dagger(\boldsymbol{r}) \end{bmatrix} = \sum_\nu \begin{bmatrix} u_\nu(\boldsymbol{r}) & v_\nu^*(\boldsymbol{r}) \\ v_\nu(\boldsymbol{r}) & u_\nu^*(\boldsymbol{r}) \end{bmatrix} \begin{bmatrix} \gamma_\nu \\ \gamma_\nu^\dagger \end{bmatrix}. \tag{5.76}$$

Here, we omit spin $\uparrow$ from both the operators and the wave functions.

We first describe the Andreev bound states at a surface of a spin-triplet p_x-wave superconductor. By repeating the argument in Sect. 5.3, it is possible to confirm the presence of the Andreev bound states at a surface of a p_x-wave superconductor in Fig. 5.6a. The wave function of the right-going wave in the superconductor is given by

$$\phi_S(\boldsymbol{r}) = \left[\begin{bmatrix} 1 \\ \Gamma_+ \end{bmatrix} e^{ik_x x} C + \begin{bmatrix} \Gamma_- \\ 1 \end{bmatrix} e^{-ik_x x} D\right] e^{-x/\xi_0} \sqrt{\frac{2}{W}} \sin\left(\frac{\pi l y}{W}\right), \tag{5.77}$$

$$\Gamma_\pm = \frac{\Delta_\pm}{E + \sqrt{E^2 - \Delta_\pm^2}}, \quad \Delta_\pm = \pm\Delta_0 \frac{k_x}{k_F}, \tag{5.78}$$

where we consider surface bound states localized at $x = 0$ by assuming $E < \Delta_0$ and we apply the hard-wall boundary condition in the y-direction. From the boundary condition $\phi_S(\boldsymbol{r})|_{x=0} = 0$, $\Gamma_+\Gamma_- = 1$ is necessary to have the surface bound state. We find that a bound state stays at $E = 0$ when

$$s_+ s_- = -1 \tag{5.79}$$

is satisfied. The wave function of the bound state can be described by

$$\phi_{E=0,l}(\boldsymbol{r}) = \begin{pmatrix} e^{i\pi/4} \\ e^{-i\pi/4} \end{pmatrix} f_l(\boldsymbol{r}), \tag{5.80}$$

$$f_l(\boldsymbol{r}) = A \sin(k_x x)\, e^{-x/\xi_0} \sqrt{\frac{2}{W}} \sin\left(\frac{\pi l y}{W}\right), \tag{5.81}$$

where A is a real constant for normalizing the wave function and $l = 1, 2, \cdots N_c$ indicates the zero-energy states. The Bogoliubov transformation in Eq. (5.76) connects the operator of the surface state and the operator of an electron at each propagating channel l,

$$\begin{bmatrix} \psi_l(\boldsymbol{r}) \\ \psi_l^\dagger(\boldsymbol{r}) \end{bmatrix} = f_l(\boldsymbol{r}) \begin{bmatrix} e^{i\pi/4} & e^{i\pi/4} \\ e^{-i\pi/4} & e^{-i\pi/4} \end{bmatrix} \begin{bmatrix} \gamma_l \\ \gamma_l^\dagger \end{bmatrix}. \tag{5.82}$$

Equation (5.82) is represented as

$$\psi_l(\boldsymbol{r}) = e^{i\pi/4}\gamma_l(\boldsymbol{r}), \quad \psi_l^\dagger(\boldsymbol{r}) = e^{-i\pi/4}\gamma_l(\boldsymbol{r}), \tag{5.83}$$

by using an fermion operator

$$\gamma_l(\boldsymbol{r}) \equiv f_l(\boldsymbol{r})(\gamma_l + \gamma_l^\dagger). \tag{5.84}$$

We find the relation

$$\gamma_l(\boldsymbol{r}) = \gamma_l^\dagger(\boldsymbol{r}), \tag{5.85}$$

which is called the Majorana relation. An spin-triplet p_x-wave superconductor hosts the Majorana Fermions at its surface (Kitaev 2001). The presence of more than one zero-energy state is responsible for the anomalous proximity effect of a spin-triplet superconductor (Tanaka and Kashiwaya 2004; Asano et al. 2006; Ikegaya et al. 2016) as we will discuss in Chap. 7.

To define the Brillouin zone clearly, we describe the BdG equation on a two-dimensional tight-binding lattice as

$$\sum_{\boldsymbol{r}'} \begin{bmatrix} \xi(\boldsymbol{r}, \boldsymbol{r}') & \Delta(\boldsymbol{r}, \boldsymbol{r}') \\ -\Delta^*(\boldsymbol{r}, \boldsymbol{r}') & -\xi^*(\boldsymbol{r}, \boldsymbol{r}') \end{bmatrix} \begin{bmatrix} u(\boldsymbol{r}') \\ v(\boldsymbol{r}') \end{bmatrix} = E \begin{bmatrix} u(\boldsymbol{r}) \\ v(\boldsymbol{r}) \end{bmatrix}, \tag{5.86}$$

$$\xi(\boldsymbol{r}, \boldsymbol{r}') = -t\left(\delta_{\boldsymbol{r},\boldsymbol{r}'+\hat{x}} + \delta_{\boldsymbol{r},\boldsymbol{r}'-\hat{x}} + \delta_{\boldsymbol{r},\boldsymbol{r}'+\hat{y}} + \delta_{\boldsymbol{r},\boldsymbol{r}'-\hat{y}}\right) + v_{\boldsymbol{r}}\delta_{\boldsymbol{r},\boldsymbol{r}'} - \epsilon_F \delta_{\boldsymbol{r},\boldsymbol{r}'}, \tag{5.87}$$

$$\Delta(\boldsymbol{r}, \boldsymbol{r}') = i\frac{\Delta_0}{2}\left(\delta_{\boldsymbol{r},\boldsymbol{r}'+\hat{x}} - \delta_{\boldsymbol{r},\boldsymbol{r}'-\hat{x}}\right), \quad \boldsymbol{r} = j\hat{\boldsymbol{x}} + m\hat{\boldsymbol{y}}, \tag{5.88}$$

where t is the hopping integral between the nearest neighbor lattice sites, and the on-site potential $v_{\boldsymbol{r}}$ is set to be zero. When we apply the periodic boundary condition in both the x- and the y-directions, the wave function is represented in the Fourier series as

$$u(\boldsymbol{r}) = \sum_{k_x, k_y} u_{k_x,k_y} \sqrt{\frac{1}{L}} e^{ik_x j} \sqrt{\frac{1}{M}} e^{ik_y m}, \quad k_x = 2\pi n/L, \quad k_y = 2\pi l/M, \tag{5.89}$$

with $-L/2 < n \leq L/2$ and $-M/2 < l \leq M/2$, where L and M is the number of lattice sites in the x- and y-directions, respectively. The lattice constant is set to be unity. The wave function represents the eigenstate of the BdG Hamiltonian. The energy eigenvalue is calculated as $E = \pm\sqrt{\xi_{\boldsymbol{k}}^2 + \Delta_{\boldsymbol{k}}^2}$ with

$$\xi_{\boldsymbol{k}} = -2t(\cos k_x + \cos k_y - 2) - \epsilon_F, \quad \Delta_{\boldsymbol{k}} = \Delta_0 \sin k_x. \tag{5.90}$$

In Fig. 5.6b, we plot the energy eigenvalues as a function of k_y, where we choose parameters as $\Delta_0 = t$, $\epsilon_F = 2t$, $L = 500$, and $M = 200$. The calculated results show that the pair potential has nodes on the Fermi surface at $k_y = \pm k_F$ with $k_F = \pi/2$ for the current parameter choice. The periodic boundary conditions in both the x- and the y-directions indicate that the shape of a superconductor is a two-dimensional sphere. There is no edge or surface in such a superconductor. We note that the eigenvalues at $E = 0$ between the two nodal points in Fig. 5.6b are absent under such boundary conditions. To have edges or surfaces, we need to apply the hard-wall boundary condition in the x-direction. When we introduce the hard-wall potential at $j = 0$ and $j = L$, zero-energy states appear as indicated by dots between the two nodal points in Fig. 5.6b. Namely, a surface of a p_x-wave superconductor hosts bound states at zero energy. The wave function of such surface bound states is described by Eq. (5.80).

Our remaining task is to calculate a topological number under the periodic boundary condition in the two directions. However, it is impossible to apply the naive topological classification to a p_x-wave superconductor because the superconducting gap has nodes on the Fermi surface as shown in Fig. 5.6b. Namely, the energy spectra are not gapped in the whole Brillouin zone. To topologically characterize such a nodal superconductor, we apply a theoretical prescription called dimensional reduction (Sato et al. 2011). We fix k_y at a certain point other than the nodal points $k_y = \pm\pi/2$ and consider the one-dimensional Brillouin zone of $-\pi < k_x \leq \pi$. The one-dimensional Brillouin zone in this case is identical to a sphere in one-dimension S^1 because $k_x = -\pi$ and $k_x = \pi$ indicate an identical quantum state. As we choose

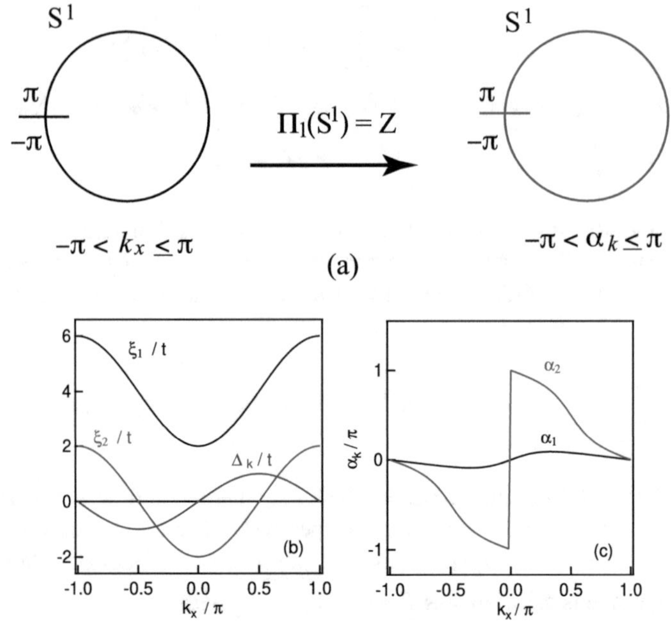

Fig. 5.7 **a** The winding number $\mathcal{Z}$ counts the number how many times $\alpha_{\boldsymbol{k}}$ encircles the one-dimensional sphere while k_x encircles the one-dimensional Brillouin zone once. **b** The dispersion of two channels is shown. ξ_1 is the dispersion of an evanescent channel and is away from the Fermi level $E = 0$. ξ_2 is the dispersion of a propagating channel and comes across the Fermi level. The pair potential $\Delta_{\boldsymbol{k}}$ is an odd function of k_x. **c** The mapping function $\alpha_{\boldsymbol{k}}$ is plotted as a function of k_x. For an evanescent channel, α_1 remains around zero, which results in $\mathcal{W} = 0$. For a propagating channel, α_2 changes by -2π while k_x moves from $-\pi$ to π

k_y away from the nodal points, the pair potential opens up the superconducting gap in energy spectra of such a one-dimensional Brillouin zone. We have already known the wave function of an occupied state

$$\psi_{\boldsymbol{k}} = \begin{bmatrix} -v_{\boldsymbol{k}} \,\mathrm{sgn}(\Delta_{\boldsymbol{k}}) \\ u_{\boldsymbol{k}} \end{bmatrix}, \quad u_{\boldsymbol{k}} = \sqrt{\frac{1}{2}\left(1 + \frac{\xi_{\boldsymbol{k}}}{E_{\boldsymbol{k}}}\right)}, \quad v_{\boldsymbol{k}} = \sqrt{\frac{1}{2}\left(1 - \frac{\xi_{\boldsymbol{k}}}{E_{\boldsymbol{k}}}\right)}, \tag{5.91}$$

with $\boldsymbol{k} = (k_x, k_y)$. An eigenstate is specified by an angle of $-\pi < \alpha_{\boldsymbol{k}} \leq \pi$ as

$$\cos(\alpha_{\boldsymbol{k}}) \equiv \frac{\xi_{\boldsymbol{k}}}{E_{\boldsymbol{k}}}, \quad \sin(\alpha_{\boldsymbol{k}}) \equiv \frac{\Delta_{\boldsymbol{k}}}{E_{\boldsymbol{k}}}, \quad \psi_{\boldsymbol{k}} = \begin{bmatrix} \sin\left(-\frac{\alpha_{\boldsymbol{k}}}{2}\right) \\ \cos\left(-\frac{\alpha_{\boldsymbol{k}}}{2}\right) \end{bmatrix}. \tag{5.92}$$

The function $\alpha_{\boldsymbol{k}}$ represents a mapping from the one-dimensional Brillouin zone $-\pi < k_x \leq \pi$ to the Hilbert space as illustrated in Fig. 5.7a.

It is possible to define the winding number by using the mapping,

$$\mathcal{W}(k_y) = \frac{1}{2\pi} \int_{-\pi}^{\pi} dk_x \, \partial_{k_x}(-\alpha_{\boldsymbol{k}}), \tag{5.93}$$

where $\mathcal{W}$ represents how many times $\alpha_{\boldsymbol{k}}$ encircles the one-dimensional sphere while k_x encircles the one-dimensional Brillouin zone once. In Fig. 5.7b, we show the dispersion in the normal state and the pair potential in Eq. (5.90) in the one-dimensional Brillouin zone. At ξ_1, we choose a k_y so that the dispersion is always above the Fermi level and the channel is an evanescent mode. On the other hand, at ξ_2, we choose a k_y so that the dispersion comes across the Fermi level and the channel becomes propagating mode. The wavenumbers $-\pi/2 < k_y < \pi/2$ in Fig. 5.6b correspond to the propagating channels and those $\pi/2 < |k_y| < \pi$ describe the evanescent channels. The pair potential $\Delta_{\boldsymbol{k}}$ is an odd function of k_x. In Fig. 5.7c, we plot the mapping function $\alpha_{\boldsymbol{k}}$ for the two transport channels. In an evanescent channel, α_1 depends on k_x only slightly. As a consequence, the winding number is zero. On the other hand in a propagating channel, α_2 changes by -2π while k_x encircles the Brillouin zone once, which results in $\mathcal{W} = 1$. According to the bulk-boundary correspondence, a zero-energy state appears at the surface of the superconductor for each propagating channel. As a result, a p_x-wave superconductor in two dimensions hosts highly degenerate surface bound states at zero energy as shown in Fig. 5.6b. The degree of the degeneracy at zero energy is equal to the number of the propagating channels N_c.

To summarize, the winding number is calculated to be

$$\mathcal{W}(k_y) = s_+ \frac{1 - s_+ s_-}{2}, \tag{5.94}$$

when k_y indicates a propagating channel away from the nodal points of the pair potential. The factor $s_\pm$ represents the sign of the two pair potentials defined in Eq. (5.30). In other words, a nodal superconductor is topologically nontrivial for $s_+ s_- = -1$ which is identical to the condition for the appearance of a surface Andreev bound state in Eq. (5.79). The results in Eq. (5.94) can be applied also to a d_{xy}-wave superconductor. As shown in Fig. 5.6c, zero-energy states appear at a surface of a d_{xy}-wave superconductor. The winding number in Eq. (5.94) is positive unity for $k_y > 0$ and is negative unity for $k_y < 0$ as a result of d_{xy}-wave symmetry.

References

Asano, Y., Tanaka, Y., Kashiwaya, S.: Phys. Rev. B **69**, 134501 (2004). https://doi.org/10.1103/PhysRevB.69.134501

Asano, Y., Tanaka, Y., Kashiwaya, S.: Phys. Rev. Lett. **96**, 097007 (2006). https://doi.org/10.1103/PhysRevLett.96.097007

Asano, Y.: Phys. Rev. B **64**, 224515 (2001). https://doi.org/10.1103/PhysRevB.64.224515

Buchholtz, L.J., Zwicknagl, G.: Phys. Rev. B **23**, 5788 (1981). https://doi.org/10.1103/PhysRevB.23.5788

Hara, J., Nagai, K.: Prog. Theoret. Phys. **76**(6), 1237 (1986). https://doi.org/10.1143/PTP.76.1237

Ikegaya, S., Suzuki, S.I., Tanaka, Y., Asano, Y.: Phys. Rev. B **94**, 054512 (2016). https://doi.org/10.1103/PhysRevB.94.054512

Kitaev, A.Y.: Physics-Uspekhi **44**(10S), 131 (2001). https://doi.org/10.1070/1063-7869/44/10s/s29

Sato, M., Tanaka, Y., Yada, K., Yokoyama, T.: Phys. Rev. B **83**, 224511 (2011). https://doi.org/10.1103/PhysRevB.83.224511

Tanaka, Y., Kashiwaya, S.: Phys. Rev. B **53**, R11957 (1996). https://doi.org/10.1103/PhysRevB.53.R11957

Tanaka, Y., Kashiwaya, S.: Phys. Rev. B **70**, 012507 (2004). https://doi.org/10.1103/PhysRevB.70.012507

Tanaka, Y., Kashiwaya, S.: Phys. Rev. Lett. **74**, 3451 (1995). https://doi.org/10.1103/PhysRevLett.74.3451

Chapter 6
Various Josephson Junctions

Abstract The Josephson current between two different superconductors is discussed. The calculated results show that the Josephson effect is sensitive to the difference in the pairing symmetries between two superconductors. Because of such severe selection rules, the Josephson junction can be a symmetry tester of a superconductor whose pairing symmetry has not been resolved yet. We also discuss the transport properties of chiral superconductors and those of a helical superconductor.

6.1 General Formula of the Josephson Current

In Sects. 4.4 and 5.4, we discussed the Josephson effect between the two superconductors which are identical to each other. The Josephson effect, however, depends sensitively on the relative symmetry difference between two superconductors. The purpose of this chapter is to demonstrate such rich properties of the Josephson current. The wave function in the two superconductors are described by

$$\begin{aligned}\phi_{\rm L}(\boldsymbol{r}) =&\check{\Phi}_{\rm L}\left[\begin{pmatrix} u_{L+}\hat{\sigma}_0 \\ \tilde{v}_{L+}\hat{\sigma}_L^\dagger \end{pmatrix} e^{ik_x x}\alpha + \begin{pmatrix} \tilde{v}_{L-}\hat{\sigma}_L \\ u_{L-}\hat{\sigma}_0 \end{pmatrix} e^{-ik_x x}\beta \right.\\ &\left. + \begin{pmatrix} u_{L-}\hat{\sigma}_0 \\ \tilde{v}_{L-}\hat{\sigma}_L^\dagger \end{pmatrix} e^{-ik_x x}A + \begin{pmatrix} \tilde{v}_{L+}\hat{\sigma}_L \\ u_{L+}\hat{\sigma}_0 \end{pmatrix} e^{ik_x x}B\right] f_{k_y}(y), \end{aligned} \tag{6.1}$$

$$\phi_{\rm R}(\boldsymbol{r}) =\check{\Phi}_{\rm R}\left[\begin{pmatrix} u_{R+}\hat{\sigma}_0 \\ \tilde{v}_{R+}\hat{\sigma}_R^\dagger \end{pmatrix} e^{ik_x x}C + \begin{pmatrix} \tilde{v}_{R-}\hat{\sigma}_R \\ u_{R-}\hat{\sigma}_0 \end{pmatrix} e^{-ik_x x}D\right] f_{k_y}(y), \tag{6.2}$$

with Eq. (5.58) and

$$u_{j\pm} =\sqrt{\frac{1}{2}\left(1+\frac{\Omega_{n,j\pm}}{\hbar\omega_n}\right)}, \quad v_{j\pm} = \sqrt{\frac{1}{2}\left(1-\frac{\Omega_{n,j\pm}}{\hbar\omega_n}\right)}, \quad \tilde{v}_{j\pm} = v_{j\pm}\, s_{j\pm} \tag{6.3}$$

$$\Delta_{j\pm} =\Delta_j(\pm k_x, k_y), \quad s_{j\pm} = \frac{\Delta_{j\pm}}{|\Delta_{j\pm}|}, \quad \Omega_{n,j\pm} = \sqrt{\hbar^2\omega_n^2 + \Delta_{j\pm}^2}, \tag{6.4}$$

Y. Asano, *Andreev Reflection in Superconducting Junctions*,
SpringerBriefs in Physics, https://doi.org/10.1007/978-981-16-4165-7_6

for $j = L$ or R. The pair potentials are described as

$$\hat{\Delta}_{j\pm} = \hat{\sigma}_j\, \Delta_{j\pm}, \quad \hat{\sigma}_j = \begin{cases} i\hat{\sigma}_2 & : \text{singlet} \\ i\boldsymbol{d}_j \cdot \hat{\boldsymbol{\sigma}}\hat{\sigma}_2 & : \text{triplet} \end{cases}. \tag{6.5}$$

By substituting the wave functions into the boundary conditions in Eqs. (4.31) and (4.32), it is possible to derive the relations among the amplitudes of wave functions,

$$\check{U}_R \begin{pmatrix} C \\ D \end{pmatrix} = \check{\Phi}\left[\check{U}_{L1}\begin{pmatrix} \alpha \\ \beta \end{pmatrix} + \check{U}_{L2}\begin{pmatrix} A \\ B \end{pmatrix}\right], \tag{6.6}$$

$$\check{U}_R\, \bar{k}\, \check{T}_3 \begin{pmatrix} C \\ D \end{pmatrix} = \check{\Phi}\left[\check{U}_{L1}(\bar{k}\,\check{T}_3 - 2iz_0)\begin{pmatrix} \alpha \\ \beta \end{pmatrix} + \check{U}_{L2}(-\bar{k}\,\check{T}_3 - 2iz_0)\begin{pmatrix} A \\ B \end{pmatrix}\right], \tag{6.7}$$

$$\check{U}_{L1} = \begin{pmatrix} u_{L+}\hat{\sigma}_0 & \tilde{v}_{L-}\hat{\sigma}_L \\ \tilde{v}_{L+}\hat{\sigma}_L^\dagger & u_{L-}\hat{\sigma}_0 \end{pmatrix}, \quad \check{U}_{L2} = \begin{pmatrix} u_{L-}\hat{\sigma}_0 & \tilde{v}_{L+}\hat{\sigma}_L \\ \tilde{v}_{L-}\hat{\sigma}_L^\dagger & u_{L+}\hat{\sigma}_0 \end{pmatrix}, \tag{6.8}$$

$$\check{U}_R = \begin{pmatrix} u_{R+}\hat{\sigma}_0 & \tilde{v}_{R-}\hat{\sigma}_R \\ \tilde{v}_{R+}\hat{\sigma}_R^\dagger & u_{R-}\hat{\sigma}_0 \end{pmatrix}, \quad \check{T}_3 = \begin{pmatrix} \hat{\sigma}_0 & 0 \\ 0 & -\hat{\sigma}_0 \end{pmatrix}, \quad \check{\Phi} = \check{\Phi}_{\mathrm{R}}^*\, \check{\Phi}_{\mathrm{L}}. \tag{6.9}$$

By eliminating C and D, we find

$$\begin{pmatrix} A \\ B \end{pmatrix} = -\check{Y}_{AB}^{-1}\check{Y}_{ab}\begin{pmatrix} \alpha \\ \beta \end{pmatrix} = \begin{pmatrix} \hat{r}_{ee} & \hat{r}_{eh} \\ \hat{r}_{he} & \hat{r}_{hh} \end{pmatrix}\begin{pmatrix} \alpha \\ \beta \end{pmatrix} \tag{6.10}$$

$$\check{Y}_{ab} = \left[\frac{\bar{k}}{2}(\check{T}_3\check{V}_1 - \check{V}_1\check{T}_3) + iz_0\check{V}_1\right], \quad \check{Y}_{AB} = \left[\frac{\bar{k}}{2}(\check{T}_3\check{V}_2 + \check{V}_2\check{T}_3) + iz_0\check{V}_2\right], \tag{6.11}$$

$$\check{V}_1 = \check{U}_R^{-1}\,\check{\Phi}\,\check{U}_{L1} = \begin{pmatrix} \hat{a}_1 & \hat{b}_1 \\ \hat{c}_1 & \hat{d}_1 \end{pmatrix}, \quad \check{V}_2 = \check{U}_R^{-1}\,\check{\Phi}\,\check{U}_{L2} = \begin{pmatrix} \hat{a}_2 & \hat{b}_2 \\ \hat{c}_2 & \hat{d}_2 \end{pmatrix}. \tag{6.12}$$

The Andreev reflection coefficients can be calculated as

$$\hat{r}_{\mathrm{he}} = \left[-\hat{c}_2^{-1}\hat{d}_2 + |r_n|^2\hat{a}_2^{-1}\hat{b}_2\right]^{-1}\left[\hat{c}_2^{-1}\hat{c}_1 - |r_n|^2\hat{a}_2^{-1}\hat{a}_1\right], \tag{6.13}$$

$$\hat{r}_{\mathrm{eh}} = \left[-\hat{b}_2^{-1}\hat{a}_2 + |r_n|^2\hat{d}_2^{-1}\hat{c}_2\right]^{-1}\left[\hat{b}_2^{-1}\hat{b}_1 - |r_n|^2\hat{d}_2^{-1}\hat{d}_1\right]. \tag{6.14}$$

The electric Josephson current is calculated based on a general formula, (Asano 2001),

$$J = \frac{e}{2\hbar}k_B T\sum_{\omega_n}\sum_{k_y}\mathrm{Tr}\left[\frac{\hat{\Delta}_{L+}\hat{r}_{\mathrm{he}}}{\Omega_{n,L+}} - \frac{\hat{\Delta}_{L-}^\dagger\hat{r}_{\mathrm{eh}}}{\Omega_{n,L-}}\right]. \tag{6.15}$$

6.2 Singlet-Singlet Junction

Let us consider Josephson junctions consisting of two spin-singlet superconductors. The results of the Andreev reflection coefficients are summarized as

$$\begin{aligned}\hat{r}_{\text{he}} =& \frac{(i\hat{\sigma}_2)^\dagger}{\Xi} \left[(1 - \Gamma_{R+}\Gamma_{R-})(\Gamma_{L-} - \Gamma_{L+}) \right. \\ &\left. + |t_n|^2 \left\{ \Gamma_{R+}\chi^2 + \Gamma_{R-}\Gamma_{L+}\Gamma_{L-}(\chi^*)^2 - \Gamma_{R+}\Gamma_{R-}\Gamma_{L+} - \Gamma_{L-} \right\} \right], \quad (6.16)\\ \hat{r}_{\text{eh}} =& \frac{i\hat{\sigma}_2}{\Xi} \left[(1 - \Gamma_{R+}\Gamma_{R-})(\Gamma_{L+} - \Gamma_{L-}) \right. \\ &\left. + |t_n|^2 \left\{ \Gamma_{R-}(\chi^*)^2 + \Gamma_{R+}\Gamma_{L+}\Gamma_{L-}\chi^2 - \Gamma_{R+}\Gamma_{R-}\Gamma_{L-} - \Gamma_{L+} \right\} \right], \quad (6.17)\\ \Xi(\varphi) =& (1 - \Gamma_{R+}\Gamma_{R-})(1 - \Gamma_{L+}\Gamma_{L-}) \\ &+ |t_n|^2 \left[\Gamma_{R+}\Gamma_{R-} + \Gamma_{L+}\Gamma_{L-} - \Gamma_{R+}\Gamma_{L+}\chi^2 - \Gamma_{R-}\Gamma_{L-}(\chi^*)^2 \right], \quad (6.18)\\ \Gamma_{j\pm} =& \frac{-i\Delta_{j\pm}}{\hbar\omega_n + \Omega_{n,j\pm}}, \quad \chi = \exp\left[i\frac{\varphi_L - \varphi_R}{2} \right]. \quad (6.19)\end{aligned}$$

The expression above can be applied to any singlet-singlet junctions.

We first focus on a junction where an s-wave superconductor stays on the left and a d_{xy}-wave superconductor stays on the right. We also assume relations $\Delta_L = \Delta$ and $\Delta_R = \Delta \text{sgn}(k_x k_y)$ for simplicity. The Andreev reflection coefficients in such case are calculated to be

$$\hat{r}_{\text{he}} = \frac{(i\hat{\sigma}_2)^\dagger |t_n|^2}{i} \frac{\Delta_R(\hbar\omega_n i \sin\varphi + \Omega\cos\varphi) - \Delta_L\Omega}{2\hbar\omega_n\Omega + |t_n|^2\Delta_L\,\Delta_R\, i\sin\varphi}, \quad (6.20)$$

$$\hat{r}_{\text{eh}} = \frac{i\hat{\sigma}_2 |t_n|^2}{i} \frac{\Delta_R(\hbar\omega_n i \sin\varphi - \Omega\cos\varphi) - \Delta_L\Omega}{2\hbar\omega_n\Omega + |t_n|^2\Delta_L\,\Delta_R\, i\sin\varphi}. \quad (6.21)$$

The Josephson current results in

$$J = -\frac{e}{\hbar}\sum_{k_y} \frac{|t_n|^4\Delta^2\sin(2\varphi)}{4\sqrt{1-a^2}} \left[\frac{1}{D_{s-}}\tanh\left(\frac{D_{s-}}{2k_BT}\right) - \frac{1}{D_{s+}}\tanh\left(\frac{D_{s+}}{2k_BT}\right) \right], \quad (6.22)$$

$$D_{s\pm} = \Delta\left[\frac{\sqrt{1+a} \pm \sqrt{1-a}}{2} \right], \quad a = |t_n|^2|\sin\varphi|. \quad (6.23)$$

At a temperature near T_c, the current is approximately represented as

$$J \approx -\frac{1}{12}\frac{e\Delta}{\hbar}\left(\frac{\Delta}{2k_BT}\right)^3 \sum_{k_y} |t_n|^4 \sin(2\varphi). \quad (6.24)$$

The lowest harmonic J_1 in Eq. (4.79) is missing, (Tanaka 1994). Phenomenological description of J_1 term

$$J_1 \propto \sum_{k_y} \mathrm{Tr}\left[\hat{\Delta}_L \hat{\Delta}_R^\dagger\right] = \sum_{k_y} \Delta_L(\boldsymbol{k})\, \Delta_R(\boldsymbol{k})\, \mathrm{Tr}\left[\hat{\sigma}_L\, \hat{\sigma}_R^\dagger\right] \tag{6.25}$$

represents the selection rule of the Josephson current because it depends sensitively on the relative pairing symmetries in the two superconductors. In the present case, an s-wave pair potential is independent of k_y, whereas a d_{xy}-wave pair potential is an odd function of k_y. Therefore, J_1 vanishes as a result of the summation over propagating channels. Namely, s-wave and d_{xy}-wave symmetries do not meet the selection rule for the orbital part. We also note that $J_2 < 0$ holds in most of the Josephson junctions. At $T = 0$, the current becomes

$$J \approx -\frac{e\Delta}{\hbar} \sum_{k_y} |t_n|^2 \frac{1}{\sqrt{1 + |t_n|^2 |\sin\varphi|}} \cos(\varphi)\mathrm{sgn}(\varphi). \tag{6.26}$$

Figure 6.1a shows the Josephson critical current versus temperature. The results show the low-temperature anomaly because the Andreev bound states exist at the surface of a dxy-wave superconductor. The current-phase relationship is shown in Fig. 6.1b for several choices of temperature. The results also indicate an unusual current-phase relationship at a low temperature.

The Josephson effect in unconventional superconductor depends sensitively on relative configurations of the two pair potentials. In Fig. 6.2, the two d-wave potentials are oriented by β in the opposite manner to each other. The pair potentials are represented by

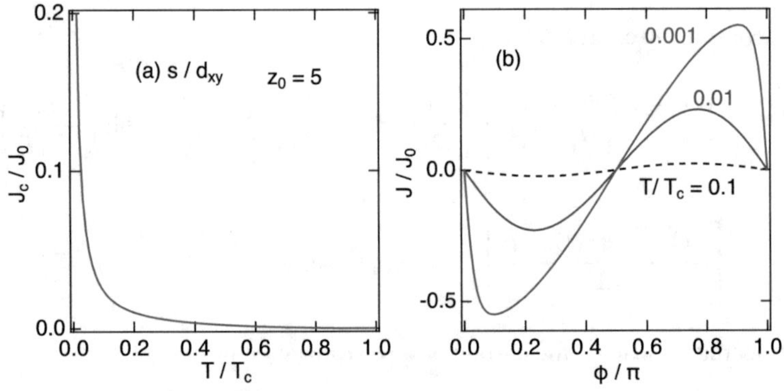

Fig. 6.1 The Josephson critical current is plotted as a function of temperature for a junction consisting of a s-wave superconductor and a d_{xy}-wave superconductor in (**a**). The current-phase relationship is shown for several choices of temperatures in (**b**)

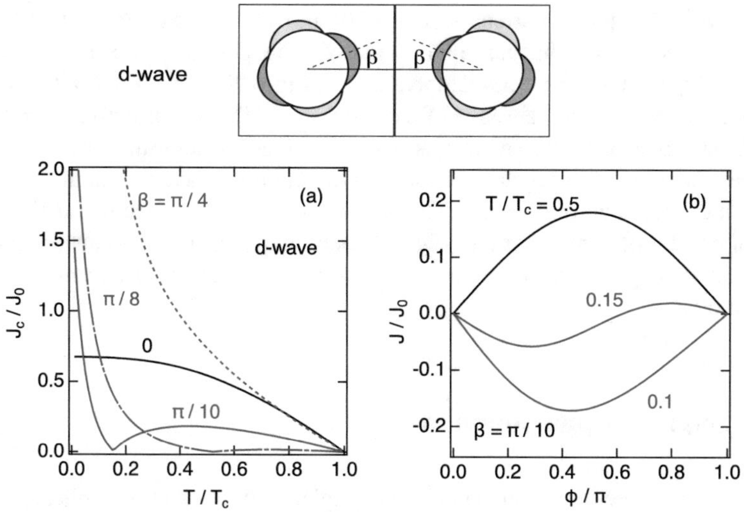

Fig. 6.2 The pair potentials in the two superconductors are oriented by β in the opposite manner to each other. The Josephson critical current is plotted as a function of temperature for several choices of β. The current-phase relationship at $\beta = \pi/10$ is shown for several temperatures in (**b**). The results at $T = 0.15T_c$ is amplified by five for better visibility

$$\Delta_L = \Delta(\cos 2\theta \cos 2\beta + \sin 2\theta \sin 2\beta), \tag{6.27}$$

$$\Delta_R = \Delta(\cos 2\theta \cos 2\beta - \sin 2\theta \sin 2\beta), \tag{6.28}$$

where $k_x = k_F \cos\theta$ and $k_y = k_F \sin\theta$. The component of $\Delta \cos 2\theta$ belongs to $d_{x^2-y^2}$-wave symmetry. On the other hand, $\Delta \sin 2\theta$ component belongs to d_{xy}-wave symmetry and causes the low-temperature anomaly. In addition, the d_{xy}-wave components in the two superconductors have the opposite sign to each other. The maximum amplitude of the Josephson current is plotted as a function of temperature for several choices of β in Fig. 6.2a. At $\beta = 0$, the Josephson current shows saturates at a low temperature. The results are identical to those with $d_{x^2-y^2}$-wave symmetry in Fig. 5.4a in Sect. 5.4. The Andreev bound states are absent at a surface of two superconductors because the relation $s_+s_- = 1$ holds for all propagating channels. At $\beta = \pi/4$, the Josephson current shows the low-temperature anomaly as a result of the relation $s_+s_- = -1$ for all propagating channels as we already discussed the results for d_{xy}-wave symmetry in Fig. 5.4a. The results for intermediate angles indicate the nonmonotonic dependence on temperature as shown in $\beta = \pi/10$ and $\pi/8$ (Tanaka and Kashiwaya 1997). The $d_{x^2-y^2}$-wave component of two pair potentials in Eq. (6.28) has the same sign in Δ_L and Δ_R, which causes the Josephson current $J_{d_{x^2-y^2}} \sin\varphi$. On the other hand, the d_{xy}-wave component in Δ_L has the opposite sign to that in Δ_R, which generates the Josephson current $-J_{d_{xy}} \sin\varphi$. The amplitude of $J_{d_{x^2-y^2}}$ saturates at a low temperature, whereas $J_{d_{xy}}$ increases rapidly with at a low temperature. As a result, $J_1 = J_{d_{x^2-y^2}} - J_{d_{xy}}$ vanishes at some temperature. In

Fig. 6.2a, the Josephson current at $\beta = \pi/10$ almost vanishes around $T = 0.15T_c$, which suggests $J_1 = 0$ happens at $T = 0.15T_c$. The current-phase relationship for $\beta = \pi/10$ is plotted for several temperatures in Fig. 6.2b. The results at $T = 0.15T_c$ shows $J \propto -\sin 2\varphi$ and absence of J_1. At $T = 0.5T_c$, the current-phase relationship is sinusoidal as usual. The results at $T = 0.1T_c$ is also sinusoidal but the current changes its sign. Such junction is called π-junction because the minimum of the junction energy stays at $\varphi = \pm\pi$. The nonmonotonic dependence of the Josephson critical current on temperature in Fig. 6.2a is a unique character to unconventional superconductor junctions, where two pair potentials are oriented as the mirror image of each other.

6.3 Triplet-Triplet Junction

The Andreev reflection coefficients in spin-triplet junctions have complicated structures due to the spin degree of freedom of a Cooper pair. It is not easy to obtain a simple analytical expression of the Andreev reflection coefficients for general cases. When $\boldsymbol{d}$ vectors in the two superconductors are identical to each other (i.e., $\boldsymbol{d}_L = \boldsymbol{d}_R$), the expression in Eqs. (6.16)–(6.18) can be applied in such triplet-triplet junctions by changing $i\hat{\sigma}_2 \rightarrow i\boldsymbol{d} \cdot \hat{\boldsymbol{\sigma}}\hat{\sigma}_2$. In such situation, the nonmonotonic dependence of the Josephson current on temperature discussed in Fig. 6.2 can be seen also in p-wave SIS junctions. The results are demonstrated in Fig. 6.3. The characteristic features in p-wave mirror junctions are essentially the same as those in d-wave mirror junctions.

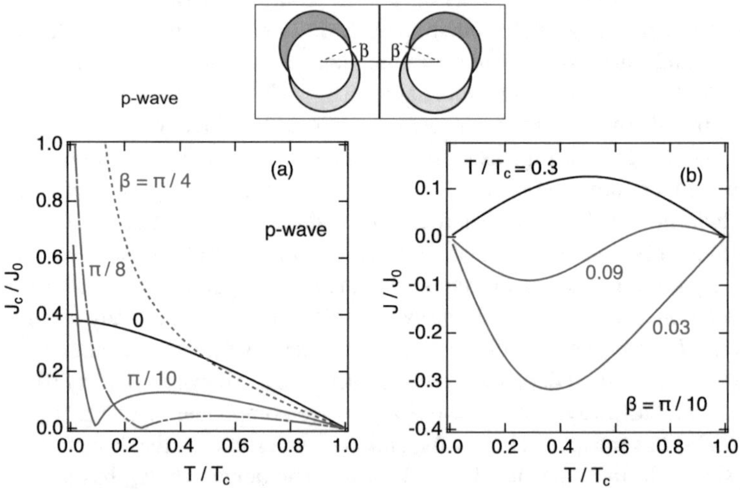

Fig. 6.3 The results of the mirror type of junction for spin-triplet p-wave symmetry. The Josephson critical current versus temperature (**a**) and the current-phase relationship at $\beta = \pi/10$ in (**b**). The results at $T = 0.09T_c$ is amplified by ten for better visibility

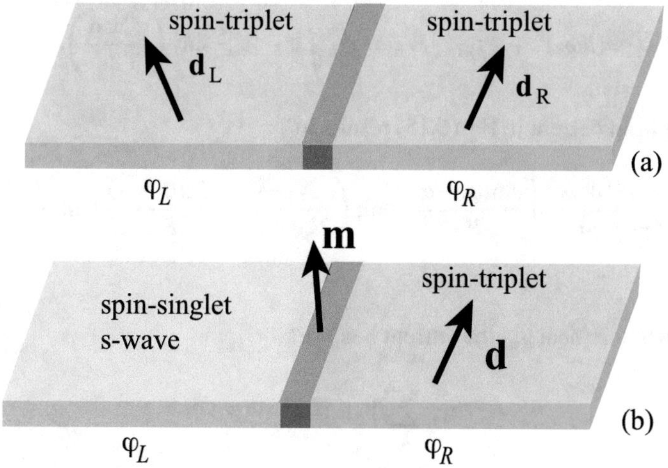

Fig. 6.4 **a** A Josephson junction consisting of two spin-triplet superconductors. An angle between the two $\boldsymbol{d}$ vectors is denoted by α. **b** A Josephson junction consisting of a spin-singlet s-wave superconductor and a spin-triplet superconductor, where $\boldsymbol{m}$ represents the magnetic moment at the insulating barrier. An angle between $\boldsymbol{d}$ vector in a spin-triplet superconductor and $\boldsymbol{m}$ is denoted by β

To discuss the effects of the relative spin configuration of two superconductors on the current, we assume that the orbital part of the pair potential $\Delta_{\boldsymbol{k}}$ is common in the two superconductors and satisfies Eq. (5.36). The pair potentials are represented as

$$\hat{\Delta}_{L+} = s_+|\Delta_{\boldsymbol{k}}| i\boldsymbol{d}_L \cdot \hat{\boldsymbol{\sigma}}\hat{\sigma}_2, \quad \hat{\Delta}_{L-} = s_-|\Delta_{\boldsymbol{k}}| i\boldsymbol{d}_L \cdot \hat{\boldsymbol{\sigma}}\hat{\sigma}_2, \tag{6.29}$$

$$\hat{\Delta}_{R+} = s_+|\Delta_{\boldsymbol{k}}| i\boldsymbol{d}_R \cdot \hat{\boldsymbol{\sigma}}\hat{\sigma}_2, \quad \hat{\Delta}_{R-} = s_-|\Delta_{\boldsymbol{k}}| i\boldsymbol{d}_R \cdot \hat{\boldsymbol{\sigma}}\hat{\sigma}_2, \tag{6.30}$$

with $s_+s_- = 1$ or -1. The $\boldsymbol{d}$ vectors are oriented from each other as shown in Fig. 6.4a and they are normalized as $|\boldsymbol{d}_L| = |\boldsymbol{d}_R| = 1$. The reflection coefficients for $s_+s_- = 1$ are calculated as

$$\hat{r}_{\mathrm{he}} = \frac{|t_n|^2 \hat{\Delta}_{L+}^{\dagger}}{2i\,\Xi_{P+}\Xi_{P-}} \big[\Omega^2(X\cos\alpha - \hbar\omega_n) + |t_n|^2 K_-(X - \hbar\omega_n\cos\alpha) + i\Omega(\Delta_{\boldsymbol{k}}^2\cos\varphi + \hbar\omega_n X - |t_n|^2 K_-)\boldsymbol{n}\cdot\hat{\boldsymbol{\sigma}}\big], \tag{6.31}$$

$$\hat{r}_{\mathrm{eh}} = \frac{|t_n|^2}{2i\,\Xi_{P+}\Xi_{P-}} \big[\Omega^2(X^*\cos\alpha - \hbar\omega_n) + |t_n|^2 K_-(X^* - \hbar\omega_n\cos\alpha) - i\Omega(\Delta_{\boldsymbol{k}}^2\cos\varphi + \hbar\omega_n X^* - |t_n|^2 K_-)\boldsymbol{n}\cdot\hat{\boldsymbol{\sigma}}^*\big]\hat{\Delta}_{L-}, \tag{6.32}$$

$$K_\pm = \frac{\Delta_{\boldsymbol{k}}^2}{2}(\cos\varphi \pm \cos\alpha), \quad X = \hbar\omega_n\cos\varphi + i\Omega_n\sin\varphi, \tag{6.33}$$

$$\boldsymbol{d}_L \cdot \boldsymbol{d}_R = \cos\alpha, \quad \boldsymbol{n} = \boldsymbol{d}_R \times \boldsymbol{d}_L, \tag{6.34}$$

$$\Xi_{P\pm} = (\hbar\omega_n)^2 + P_{t\pm}^2, \quad P_{t\pm} = \Delta_{\boldsymbol{k}}\sqrt{1 - |t_n|^2 \sin^2\left(\frac{\varphi \pm \alpha}{2}\right)}. \tag{6.35}$$

The Josephson current in Eq. (6.15) results in

$$J = \frac{e}{\hbar}\sum_{k_y} \frac{|t_n|^2 \Delta_{\boldsymbol{k}}^2}{4} \left[\frac{\sin(\varphi+\alpha)}{P_{t+}} \tanh\left(\frac{P_{t+}}{2k_B T}\right) + \frac{\sin(\varphi-\alpha)}{P_{t-}} \tanh\left(\frac{P_{t-}}{2k_B T}\right)\right]. \tag{6.36}$$

At a temperature near T_c, the current becomes

$$J \approx \frac{e}{2\hbar}\sum_{k_y} |t_n|^2 \, |\Delta_{\boldsymbol{k}}| \, \sin\varphi \, \cos\alpha. \tag{6.37}$$

The relation of $\cos\alpha = \boldsymbol{d}_L \cdot \boldsymbol{d}_R$ represents a selection rule of the Josephson charge current in spin space. Indeed, we find in Eq. (6.25) that

$$\mathrm{Tr}\left[\hat{\sigma}_L \, \hat{\sigma}_R^\dagger\right] = 2\cos\alpha. \tag{6.38}$$

The first-order coupling vanishes when the pairing functions in the two superconductors are orthogonal to each other in spin space. Since $s_+ s_- = 1$, the Josephson current is saturate at a low temperature well below T_c. The current-phase relationship of the Josephson charge current in Eq. (6.36) is plotted for several α in Fig. 6.5a, where the temperature is fixed at $T = 0.1T_c$ and the transmission probability of the barrier is chosen as 0.1. The amplitude of the current decreases with the increase of α from 0. At $\alpha = \pi/2$, the lowest order coupling vanishes as indicated by Eq. (6.37). For $\alpha > \pi/2$, the junction becomes π-junction. Flipping the direction of $\boldsymbol{d}$ vector in spin space is compensated by the π-phase shift in gauge space.

The Andreev reflection coefficients have a term proportional to $\boldsymbol{n} \cdot \hat{\boldsymbol{\sigma}}$, which is a characteristic feature of spin-triplet superconducting junctions. These reflection terms represent the flow of spin supercurrents. The spin current flowing through the junction can be calculated based on the formula, (Asano 2005, 2006),

$$\boldsymbol{J}_s = -\frac{1}{8} k_B T \sum_{\omega_n} \sum_{k_y} \mathrm{Tr}\left[\frac{\hat{\Delta}_{L+} \hat{r}_{\mathrm{he}} \hat{\boldsymbol{\sigma}} + \hat{r}_{\mathrm{he}} \hat{\Delta}_{L+} \hat{\boldsymbol{\sigma}}^*}{\Omega_{L+}} - \frac{\hat{\Delta}_{L-}^\dagger \hat{r}_{\mathrm{eh}} \hat{\boldsymbol{\sigma}}^* + \hat{r}_{\mathrm{eh}} \hat{\Delta}_{L-}^\dagger \hat{\boldsymbol{\sigma}}}{\Omega_{L-}}\right]. \tag{6.39}$$

We recover the formula for the electric current in Eq. (6.15) by replacing

$$\frac{\hbar}{2}\hat{\boldsymbol{\sigma}} \to -e\hat{\sigma}_0, \tag{6.40}$$

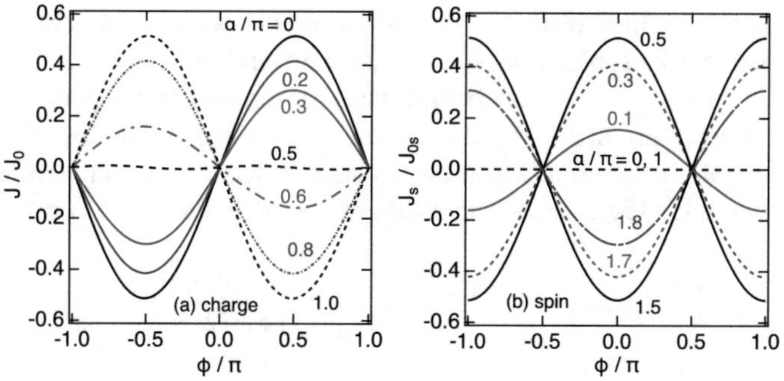

Fig. 6.5 The Josephson charge current (**a**) and spin current (**b**) in spin-triplet SIS junctions, where α is the relative angle between the two $\boldsymbol{d}$ vectors in the two superconductors. The current is calculated at $T = 0.1T_c$. We choose the transmission probability of the junction interface at 0.1

in Eq. (6.39). By substituting the Andreev reflection coefficients in Eqs. (6.31) and (6.32), the spin current is represented by

$$\boldsymbol{J}_s = \frac{\boldsymbol{n}}{8}\sum_{k_y} |t_n|^2 \Delta_{\boldsymbol{k}}^2 \left[\frac{\sin(\varphi+\alpha)}{P_{t+}} \tanh\left(\frac{P_{t+}}{2k_BT}\right) - \frac{\sin(\varphi-\alpha)}{P_{t-}} \tanh\left(\frac{P_{t-}}{2k_BT}\right)\right]. \tag{6.41}$$

At $T \lesssim T_c$, the results become

$$\boldsymbol{J}_s \propto \boldsymbol{n} \cos\varphi \, \sin\alpha. \tag{6.42}$$

The spin supercurrent polarized in $\boldsymbol{n} = \boldsymbol{d}_R \times \boldsymbol{d}_L$ direction in spin space flows across the junction, which is a unique property to spin-triplet superconductor junctions. Figure 6.5b shows the spin current in Eq. (6.41) plotted as a function of φ for several α. The spin supercurrent flows even at zero phase difference $\varphi = 0$. The spatial gradient in the $\boldsymbol{d}$ vector generates the spin supercurrent.

For $s_+ s_- = -1$, the Andreev reflection coefficients are calculated as

$$\begin{aligned}\hat{r}_{\text{he}} =& \frac{|t_n|^2 \hat{\Delta}_{L+}^{\dagger}}{2i\,\Xi_{M+}\Xi_{M-}} \Big[2(\hbar\omega_n)^3 + |t_n|^2\hbar\omega_n \left\{\Delta_{\boldsymbol{k}}^2(1+\cos\varphi\cos\alpha) - (\hbar\omega_n)^2 - \hbar\omega_n X \cos\alpha\right\} \\ &- |t_n|^4 K_+(X + \hbar\omega_n\cos\alpha) + i|t_n|^2\Omega(\hbar\omega_n X - |t_n|^2 K_+)\boldsymbol{n}\cdot\hat{\boldsymbol{\sigma}}\Big],\end{aligned} \tag{6.43}$$

$$\begin{aligned}\hat{r}_{\text{eh}} =& \frac{|t_n|^2}{2i\,\Xi_{M+}\Xi_{M-}} \Big[2(\hbar\omega_n)^3 + |t_n|^2\hbar\omega_n \left\{\Delta_{\boldsymbol{k}}^2(1+\cos\varphi\cos\alpha) - (\hbar\omega_n)^2 - \hbar\omega_n X^* \cos\alpha\right\} \\ &- |t_n|^4 K_+(X^* + \hbar\omega_n\cos\alpha) - i|t_n|^2\Omega(\hbar\omega_n X^* - |t_n|^2 K_+)\boldsymbol{n}\cdot\hat{\boldsymbol{\sigma}}^*\Big]\hat{\Delta}_{L-},\end{aligned} \tag{6.44}$$

$$\Xi_{M\pm} = (\hbar\omega_n)^2 + M_{t\pm}^2, \quad M_{t\pm} = |t_n|\Delta_{\boldsymbol{k}} \cos\left(\frac{\varphi\pm\alpha}{2}\right). \tag{6.45}$$

We only supply the final results of the Josephson electric current and spin current which are obtained by substituting $P_{t\pm}$ in Eqs. (6.36) and (6.41) by $M_{t\pm}$. The resonant transmission through the Andreev bound states causes the low-temperature anomaly in both the charge current and the spin current (Asano 2006). The Josephson charge current at $\alpha = 0$ is proportional to $\sin(\varphi/2)$ independent of t_n at zero temperature as already discussed in Fig. 5.4a, b. The results at $\alpha = \pi/2$ at zero temperature become

$$J = \frac{e}{2\hbar} \sum_{k_y} |t_n||\Delta_k| \times \begin{cases} \sin(\varphi/2) & : |\varphi| < \pi/2 \\ -\cos(\varphi/2)\,\mathrm{sgn}(\varphi) & : \pi/2 < |\varphi| \leq \pi \end{cases}, \tag{6.46}$$

which is discontinuous at $\varphi = \pm\pi/2$. The current-phase relationship is π-periodic because the lowest order coupling is absent due to the selection rule in spin space, (i.e., $\boldsymbol{d}_L \cdot \boldsymbol{d}_R = 0$).

6.4 Singlet-Triplet Junction

The selection rule for the spin configuration in Eq. (6.25) indicates that $J_1 = 0$ for a junction consisting of a spin-singlet superconductor $\hat{\sigma}_L = i\hat{\sigma}_2$ and a spin-triplet superconductor $\hat{\sigma}_R = i\boldsymbol{d} \cdot \hat{\boldsymbol{\sigma}}\,\hat{\sigma}_2$ as shown in Fig. 6.4b. The spin of a Cooper pair is unity in a triplet superconductor, whereas it is zero in a singlet superconductor. The lowest order coupling vanishes because such two pairing states are orthogonal to each other in spin space. A magnetically active interface is necessary to have the lowest order coupling. To understand these properties, let us consider a junction consisting of spin-singlet s-wave superconductor, a spin-triplet p-wave superconductor, and a material X in between. We begin this section by analyzing an expression of the lowest order Josephson current in Eq. (4.84),

$$J = \frac{ie}{\hbar} \sum_{k_y} k_B T \sum_{\omega_n} \mathrm{Tr}\left[\hat{r}_{\mathrm{eh}}(L) \cdot \hat{t}_X^h \cdot \hat{r}_{\mathrm{he}}(R) \cdot \hat{t}_X^e - \hat{r}_{\mathrm{he}}(L) \cdot \hat{t}_X^e \cdot \hat{r}_{\mathrm{eh}}(R) \cdot \hat{t}_X^h\right]. \tag{6.47}$$

The reflection processes relating to the current are illustrated in Fig. 6.6. The Andreev reflection coefficients are given in Eq. (5.34) in an NS junction,

$$\hat{r}_{\mathrm{he}} = \Gamma_+ \hat{\sigma}_S^\dagger\, e^{-i\varphi}, \quad \hat{r}_{\mathrm{eh}} = \Gamma_- \hat{\sigma}_S,\, e^{i\varphi} \quad \Gamma_\pm = \frac{-i\Delta_\pm}{\hbar\omega_n + \Omega_\pm}. \tag{6.48}$$

Here, we assume that the interface between a superconductor and the material X is highly transparent, (i.e., $|t_n| = 1$). The reflection coefficients at the left interface are

$$\hat{r}_{\mathrm{he}}(L) = -i\frac{\Delta_s}{\hbar\omega_n + \Omega_s} (i\hat{\sigma}_2)^\dagger\, e^{-i\varphi_L}, \quad \hat{r}_{\mathrm{eh}}(L) = -i\frac{\Delta_s}{\hbar\omega_n + \Omega_s}\, i\hat{\sigma}_2\, e^{i\varphi_L}, \tag{6.49}$$

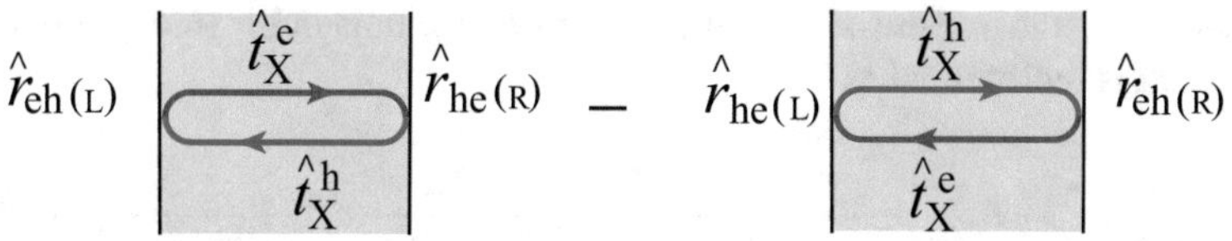

Fig. 6.6 The Andreev reflection processes carrying the Josephson current at the lowest order

where Δ_s is the amplitude of a spin-singlet s-wave superconductor on the left-hand side of a junction and $\Omega_s = \sqrt{\hbar^2\omega_n^2 + \Delta_s^2}$. At the right NS interface, the reflection coefficients can be

$$\hat{r}_{\mathrm{he}}(R) = -\, i\frac{(\Delta_+ i\boldsymbol{d}\cdot\hat{\boldsymbol{\sigma}}\hat{\sigma}_2)^\dagger}{\hbar\omega_n + \Omega_+} e^{-i\varphi_R}, \quad \hat{r}_{\mathrm{eh}}(R) = -i\frac{\Delta_- i\boldsymbol{d}\cdot\hat{\boldsymbol{\sigma}}\,\hat{\sigma}_2}{\hbar\omega_n + \Omega_-} e^{i\varphi_R}, \tag{6.50}$$

$$\Delta_\pm = \Delta(\pm k_x, k_y), \quad \Omega_\pm = \sqrt{\hbar^2\omega_n^2 + \Delta_\pm^2}. \tag{6.51}$$

In Eq. (6.47), $\hat{t}_X^e$ ($\hat{t}_X^h$) is the transmission coefficient of a material X in the electron (hole) branch. These coefficients are related to each other as

$$t_X^h = \left[t_X^e\right]^*, \tag{6.52}$$

because of particle-hole symmetry. When the transmission coefficients of X is independent of spin as $\hat{t}_X^e = t_0\hat{\sigma}_0$, it is easy to show that the lowest order coupling in Eq. (6.47) vanishes as

$$J \propto \mathrm{Tr}\left[\boldsymbol{d}\cdot\hat{\boldsymbol{\sigma}}\right] = 0, \tag{6.53}$$

as a result of the selection rule in the relative spin configuration.

Thus, the material X must be magnetically active to generate the lowest order Josephson coupling. The first example of a magnetically active layer is a ferromagnetic barrier described by

$$H_{\mathrm{FI}} = \delta(x)\left[v_0\hat{\sigma}_0 + \boldsymbol{m}\cdot\hat{\boldsymbol{\sigma}}\right], \tag{6.54}$$

where $\boldsymbol{m}$ is the magnetic moment as shown in Fig. 6.4b. The transmission coefficients of such a barrier are calculated as

$$\hat{t}_X = \frac{1}{\Xi_{\mathrm{M}}}\,\bar{k}\left[\bar{k} + iz_0 - i\boldsymbol{z}\cdot\hat{\boldsymbol{\sigma}}\right], \tag{6.55}$$

$$\hat{r}_X = \frac{1}{\Xi_{\mathrm{M}}}\left[(\bar{k} + iz_0)(-iz_0) - |\boldsymbol{z}|^2 - \bar{k}\,i\boldsymbol{z}\cdot\hat{\boldsymbol{\sigma}}\right], \tag{6.56}$$

$$z_0 = \frac{mV_0}{\hbar^2 k_F}, \quad \boldsymbol{z} = \frac{m\boldsymbol{m}}{\hbar^2 k_F}, \quad \bar{k} = k_x/k_F > 0, \quad \Xi_{\mathrm{M}} = (k + iz_0)^2 + |\boldsymbol{z}|^2. \tag{6.57}$$

By substituting the transmission coefficients to the formula in Eq. (6.47), the Josephson current is represented as

$$J = -\frac{e}{\hbar}\sum_{k_y} k_B T \sum_{\omega_n} \frac{\Delta_s}{\hbar\omega_n + \Omega_s} \frac{4\bar{k}^3 \boldsymbol{d}\cdot\boldsymbol{z}}{|\Xi_{\mathrm{M}}|^2}\left[\frac{\Delta_+}{\hbar\omega_n + \Omega_+}e^{i\varphi} - \frac{\Delta_-}{\hbar\omega_n + \Omega_-}e^{-i\varphi}\right]. \tag{6.58}$$

When a p_y-wave superconductor is on the right-hand side of the junction, we find that

$$J_{s/p_y} = 0, \tag{6.59}$$

because $\Delta_\pm$ is an odd function of k_y. The selection rule in the orbital part also affects the Josephson effect. For a p_x-wave symmetry, by considering $\Delta_- = -\Delta_+$, the current becomes

$$J_{s/p_x} = -\frac{e}{\hbar} k_B T \sum_{\omega_n>0}\sum_{k_y} \frac{16\bar{k}^3|\boldsymbol{z}|}{|\Xi_{\mathrm{M}}|^2} \frac{\Delta_s}{\hbar\omega_n + \Omega_s} \frac{\Delta_+}{\hbar\omega_n + \Omega_+} \cos\beta \cos\varphi, \tag{6.60}$$

where β is the angle between $\boldsymbol{d}$ and $\boldsymbol{m}$, (Brydon et al. 2013). The magnetic moment $\boldsymbol{m} \parallel \boldsymbol{d}$ can supply spin angular momentum ± 1 to a Cooper pair. The current-phase relationship is also unusual as the current is finite even at $\varphi = 0$. Generally speaking, the breakdown of time-reversal symmetry is a necessary condition for the flow of electric current. In this junction, the magnetic moment at the interface breaks time-reversal symmetry.

The second example of a magnetically active potential is spin-orbit coupling due to the nonmagnetic barrier potential,

$$H_{\mathrm{SO}} = \delta(x)\left[v_0\hat{\sigma}_0 + \lambda\boldsymbol{e}_x\cdot(i\nabla\times\hat{\boldsymbol{\sigma}})\right], \tag{6.61}$$

where λ represents the strength of the spin-orbit interaction. The potential gradient in the x-direction (parallel direction to the current) at the interface generates a spin-orbit interaction. Here, we consider a three-dimensional Josephson junction because the selection rule depends also on the spatial dimension. The transmission coefficient is given by

$$\hat{t}^e_X(k,\boldsymbol{p}) = \frac{\bar{k}}{\Xi_{\mathrm{SO}}}\left[\bar{k} + iz_0 - i\lambda_{\mathrm{SO}}(k_y\hat{\sigma}_3 - k_z\hat{\sigma}_2)\right], \tag{6.62}$$

$$\hat{t}^h_X(k,\boldsymbol{p}) = \frac{\bar{k}}{\Xi^*_{\mathrm{SO}}}\left[\bar{k} - iz_0 - i\lambda_{\mathrm{SO}}(k_y\hat{\sigma}_3 + k_z\hat{\sigma}_2)\right], \tag{6.63}$$

$$\Xi_{\mathrm{SO}} = (\bar{k} + iz_0)^2 + \lambda_{\mathrm{SO}}^2(k_y^2 + k_z^2), \quad \boldsymbol{p} = (k_y, k_z), \tag{6.64}$$

where λ_{SO} is the coupling constant. The current is calculated as

$$J = \frac{ie}{\hbar}\sum_{p} k_B T \sum_{\omega_n} \frac{\Delta_s}{\hbar\omega_n + \Omega_s} \frac{4\bar{k}^2 z_0 \lambda_{SO}}{|\Xi_{SO}|^2} \left[\frac{\Delta_+}{\hbar\omega_n + \Omega_+} e^{i\varphi} - \frac{\Delta_-}{\hbar\omega_n + \Omega_-} e^{-i\varphi} \right]$$
$$\times (k_y d_3 - k_z d_2). \quad (6.65)$$

We find that

$$J_{s/p_x} = 0, \quad (6.66)$$

for a junction a consisting of p_x-wave superconductor because of the selection rule for the orbital part. In the case of a p_y-wave superconductor junction, the current becomes

$$J_{s/p_y} = -\frac{e}{\hbar} k_B T \sum_{\omega_n > 0} \sum_{p} \frac{16\bar{k}^2 z_0 \lambda_{SO} k_y d_3}{|\Xi_{SO}|^2} \frac{\Delta_s}{\hbar\omega_n + \Omega_s} \frac{\Delta_+}{\hbar\omega_n + \Omega_+} \sin\varphi, \quad (6.67)$$

with $\Delta_\pm = \Delta_t \bar{k}_y$ and Δ_t being the amplitude of the spin-triplet pair potential. For a p_z-wave symmetry with $\Delta_\pm = \Delta_t \bar{k}_z$, the Josephson current results in

$$J_{s/p_z} = \frac{e}{\hbar} k_B T \sum_{\omega_n > 0} \sum_{p} \frac{16\bar{k}^2 z_0 \lambda_{SO} k_z d_2}{|\Xi_{SO}|^2} \frac{\Delta_s}{\hbar\omega_n + \Omega_s} \frac{\Delta_+}{\hbar\omega_n + \Omega_+} \sin\varphi. \quad (6.68)$$

The lowest order Josephson current exists when the pair potential of a spin-triplet superconductor has components such as $\Delta_t \bar{k}_y i\hat{\sigma}_3\hat{\sigma}_2$ and $\Delta_t \bar{k}_z i\hat{\sigma}_2\hat{\sigma}_2$, which is the selection rule in the presence of a spin-orbit interaction, (Millis et al. 1988). The current-phase relationship is always sinusoidal because spin-orbit interactions preserve time-reversal symmetry.

The lowest order Josephson coupling depends sensitively on magnetically active potentials at the interface, the direction of $\boldsymbol{d}$ vector, and the orbital symmetry of spin-triplet superconductor. Therefore, measurement of the lowest order Josephson current is a good symmetry tester for a spin-triplet superconductor.

6.5 Chiral and Helical Superconductors

The most well-established spin-triplet pairing order may be the superfluid phases of liquid ^{3}He, (Volovik 2003). The orbital part belongs to p-wave symmetry. In p-wave symmetry, two He atoms can avoid an atomic hard-core repulsive interaction and form a Cooper pair. The pressure-temperature phase diagram of superfluid ^{3}He has two phases: B-phase appearing low-temperature and low-pressure regime and A-phase appearing high-temperature and high-pressure regime. In solid state physics, spin-triplet pairs have been discussed mainly on two-dimensional electron systems in layered compounds. A chiral order parameter described by

$$\hat{\Delta}_{\boldsymbol{k}} = i\Delta \frac{k_x + i\chi k_y}{k_F} \hat{\sigma}_3 \, \hat{\sigma}_2 = \frac{\Delta}{k_F} \begin{bmatrix} 0 & k_x + i\chi k_y \\ k_x + i\chi k_y & 0 \end{bmatrix} \tag{6.69}$$

is a two-dimensional analog of ^{3}He A-phase, where $\chi = \pm 1$ is called chirality of a Cooper pair. A Cooper pair consists of two electrons with opposite spin to each other as shown in Eq. (5.17) because $\boldsymbol{d}$ points the perpendicular to the two-dimensional plane. A chiral p-wave superconductor has attracted considerable attention in condensed matter physics because it hosts a Majorana Fermion at its surface. The pair potential in Eq. (6.69) has been a promising candidate of order parameter in a ruthenate superconductor, (Maeno et al. 1994, Rice and Sigrist 1995). A helical order parameter described by

$$\hat{\Delta}_{\boldsymbol{k}} = i\Delta \left[\frac{k_x}{k_F} \hat{\sigma}_1 + \frac{k_y}{k_F} \hat{\sigma}_2 \right] \hat{\sigma}_2 = \frac{\Delta}{k_F} \begin{bmatrix} -(k_x - ik_y) & 0 \\ 0 & k_x + ik_y \end{bmatrix} \tag{6.70}$$

is a two-dimensional analog of ^{3}He B-phase. A Cooper pair with $\uparrow\uparrow$-spin has the negative chirality, whereas that with $\downarrow\downarrow$-spin belongs to the positive chirality. In this section, we discuss the transport properties of such unconventional superconductors. In two dimensions, chiral superconductors are characterized by the pair potential of

$$\hat{\Delta}_{\boldsymbol{k}} = \Delta e^{in\theta} \, \hat{\sigma}_S, \quad \hat{\sigma}_S = \begin{cases} i\hat{\sigma}_2, & n = \pm 2, \pm 4, \ldots, \\ i\boldsymbol{d} \cdot \hat{\sigma} \, \hat{\sigma}_2, & n = \pm 1, \pm 3, \ldots, \end{cases} \tag{6.71}$$

where $k_x = k_F \cos\theta$ and $k_y = k_F \sin\theta$. An integer n is a topological number called the Chern number. The pair potential is even-parity spin-singlet (odd-parity spin-triplet) when n is an even (an odd) integer number. The pair potential in Eq. (6.69) corresponds to $n = 1$.

The Andreev bound states at a surface of a chiral superconductor have very unique character. Let us consider a surface of a chiral superconductor at $x = 0$. The wave function at the surface of superconductor for $E < \Delta$ can be given by

$$\phi_{\mathrm{R}}(x, y) = \left[\begin{pmatrix} E + i\sqrt{\Delta^2 - E^2} \\ \Delta e^{-in\theta} \hat{\sigma}_S^\dagger \end{pmatrix} e^{ik_x x} C + \begin{pmatrix} E - i\sqrt{\Delta^2 - E^2} \\ \Delta (-1)^n e^{in\theta} \hat{\sigma}_S^\dagger \end{pmatrix} e^{-ik_x x} D \right] e^{ik_y y} e^{-x/\xi_0},$$

where we have taken into account the relation $\Delta_+ = \Delta e^{in\theta}$ and $\Delta_- = \Delta e^{in(\pi - \theta)}$. From the boundary condition at $x = 0$, the energies of the bound states are obtained as follows:

$$\begin{aligned} E &= \Delta \, \sin\theta & &: n = 1, \\ E &= -\Delta \, \cos(2\theta) \, \mathrm{sgn}(\sin 2\theta) & &: n = 2, \\ E &= \Delta \, \sin(3\theta) \, \mathrm{sgn}(\cos 3\theta) & &: n = 3. \end{aligned} \tag{6.72}$$

Figure 6.7 shows the dispersion of the bound states. The number of the bound states at $E = 0$ is $|n|$ according to the bulk-boundary correspondence. These bound states carry the electric current along the surface of a superconductor as schematically

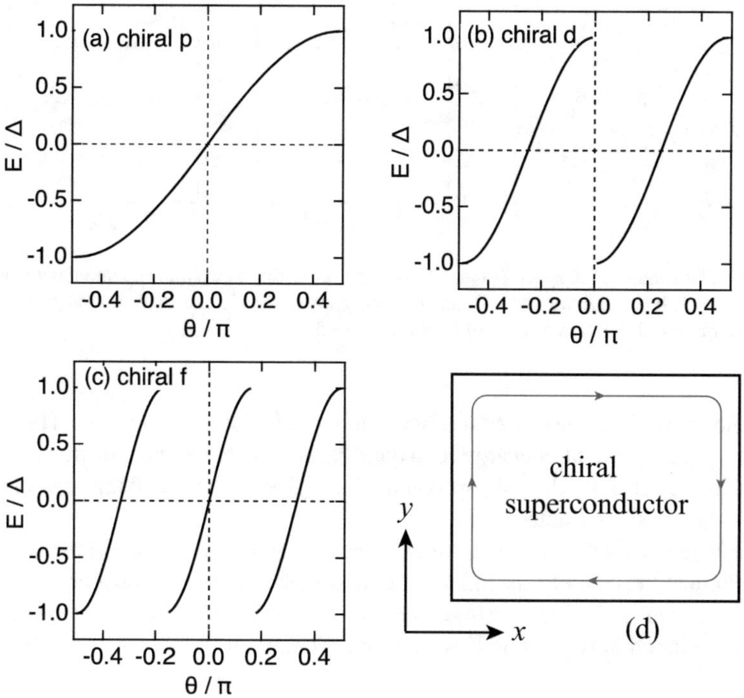

Fig. 6.7 The dispersion of the bound states at a surface of a chiral superconductor. **a** chiral p-wave $n = 1$, **b** chiral d-wave $n = 2$, and **c** chiral f-wave $n = 3$. A chiral edge mode of a chiral p-wave superconductor carries the electric current as shown in (**d**)

shown in Fig. 6.7d. When the sign of the Chern number is inverted, the dispersion changes its sign and the electric current changes its direction of flow. The tunnel spectra of a chiral superconductor can be calculated from the reflection coefficients of an NS junction,

$$\hat{r}_{\mathrm{ee}} = \frac{r_n(1-\Gamma_+\Gamma_-)}{1-|r_n|^2\Gamma_+\Gamma_-}\hat{\sigma}_0, \quad \hat{r}_{\mathrm{he}} = \frac{|t_n|^2\Gamma_+\hat{\sigma}_S^{\dagger}\, e^{-i\varphi}}{1-|r_n|^2\Gamma_+\Gamma_-}, \tag{6.73}$$

$$\Gamma_+ = \frac{\Delta_+^*}{E+\Omega}, \quad \Gamma_- = \frac{\Delta_-}{E+\Omega}, \quad \Omega = \sqrt{E^2-\Delta^2}. \tag{6.74}$$

The tunnel spectra for a chiral p-wave superconductor $n = 1$ are displayed in Fig. 6.8a. The results with $z_0 = 5$ have a dome-shaped broad peak below the gap. It may be possible to say that the subgap spectra are a result of the enhancement at the surface density of states due to a chiral edge mode. For a chiral d-wave superconductor $n = 2$, however, the tunnel spectra in Fig. 6.8b only have a shallow broad dip at the subgap energy region. The dispersion of the subgap states makes the subgap spectra complex. When we increase the Chern number more, the results for $n = 3$ in

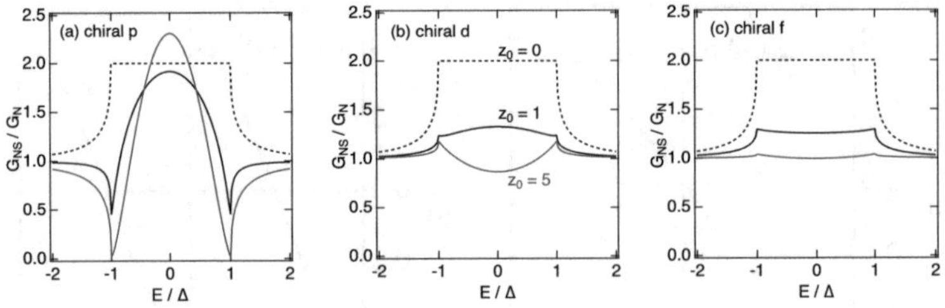

Fig. 6.8 The conductance spectra of an NS junction consisting of a chiral superconductor. **a** chiral p-wave $n = 1$, **b** chiral d-wave $n = 2$, and **c** chiral f-wave $n = 3$. The transmission probability of the barrier is about 0.026 as a result of choosing $z_0 = 5$

Fig. 6.8c show almost no specific subgap structure in the tunnel spectra. The density of states is independent of energy for a quasiparticle with the linear dispersion in one dimension. Therefore, the subgap spectra for higher n become flatter and closer to the normal density of states.

The Josephson effect depends on two Chern numbers (n and m) in the two chiral superconductors. Namely, the index theorem suggests that the number of zero-energy states is $|n - m|$ at the junction interface (Volovik 1997). The wave function on either side of the junction is given in Eqs. (6.1) and (6.2), with

$$\Delta_{L+} = \Delta e^{in\theta}, \; \Delta_{L-} = (-1)^n \Delta e^{-in\theta}, \; \Delta_{R+} = \Delta e^{im\theta}, \; \Delta_{R-} = (-1)^m \Delta e^{-im\theta}. \tag{6.75}$$

The Andreev reflection coefficients are shown in Eqs. (6.16) and (6.17) for singlet-singlet junctions. In the case of a triplet-triplet junction, $i\hat{\sigma}_2$ should be replaced by $\hat{\sigma}_S$. Here, we consider triplet-triplet junctions where both n and m are odd integers. The Josephson current is calculated in terms of the energy of the Andreev bound states $\epsilon_\pm$ as

$$\begin{aligned} J =& \frac{e\Delta}{2\hbar} \sum_{k_y} \frac{\Delta |t_n|^2 \sin\varphi}{2\sin\alpha_{nm}} \\ & \times \left[\frac{\sin(2X_+)}{\epsilon_+} \tanh\left[\frac{\epsilon_+}{2k_BT}\right] + \frac{\sin(2X_-)}{\epsilon_-} \tanh\left[\frac{\epsilon_-}{2k_BT}\right]\right], \end{aligned} \tag{6.76}$$

$$\epsilon_\pm = \mp \Delta \mathrm{sgn}[\cos(X_\pm)] \sin(X_\pm), \quad X_\pm = \frac{\alpha_{nm} \pm (n-m)\theta}{2}, \tag{6.77}$$

$$\cos\alpha_{nm} = (1 - |t_n|^2) \cos\{(n+m)\theta\} - |t_n|^2 \cos\varphi, \quad 0 \le \alpha_{nm} \le \pi. \tag{6.78}$$

From the expression of the bound state energy $\epsilon_\pm$, it is easy to count the number of Andreev bound states at zero energy. At $|t_n| = 0$, the bound state energy

$$\epsilon_+ = -\Delta \mathrm{sgn}(\cos n\theta) \sin n\theta, \quad \epsilon_- = \Delta \mathrm{sgn}(\cos m\theta) \sin m\theta \tag{6.79}$$

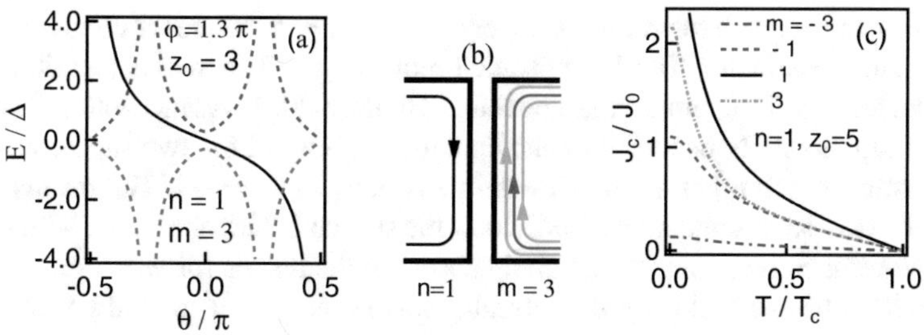

Fig. 6.9 **a** The left- and the right-hand side of Eq. (6.80) are plotted with a solid line and two broken lines, respectively. We choose $n = 1$, $m = 3$, $z_0 = 3$, and $\varphi = 1.3\pi$. **b** Schematic picture of the chiral edge currents in the two superconductors. **c** The Josephson critical current versus temperature for $z_0 = 5$. We fix the Chern number at $n = 1$ in the left superconductor. The Chern number in the right superconductor is m

represent the dispersion of the surface bound states of the two isolated superconductors. It is easy to confirm that the number of zero-energy surface states of ϵ_+ is $|n|$ and that of ϵ_- is $|m|$. At finite $|t_n|$, the solutions of $\epsilon_\pm = 0$ in Eq. (6.77) require the relation

$$\tan\left\{\frac{(n-m)\theta}{2}\right\} = \pm\tan\left(\frac{\alpha_{nm}}{2}\right). \tag{6.80}$$

The left-hand side of Eq. (6.80) goes positive infinity $|n-m|/2$ times and goes negative infinity $|n-m|/2$ times at the interval of $-\pi/2 \leq \theta \leq \pi/2$. Since $0 \leq \alpha_{nm} \leq \pi$, $\tan\left(\frac{\alpha_{nm}}{2}\right)$ is always positive. Thus, the number of the solutions of Eq. (6.80) is $|n-m|$ which corresponds to the difference in the Chern numbers between the two superconductors. In Fig. 6.9a, we plot the left-hand side of Eq. (6.80) with a solid line and the right-hand side with two broken lines, where we choose $n = 1, m = 3, z_0 = 3$, and $\varphi = 1.3\pi$. The results show that there are only two solutions, (i.e., $2 = |1-3|$). The number of the solutions is independent of the junction parameters such as z_0 and φ. As schematically illustrated in Fig. 6.9b, there are four chiral edges at $|t_n| = 0$. The chiral current in the left superconductor flows in the opposite direction to those in the right superconductor. For $|t_n| > 0$, two chiral currents flowing in the opposite direction cancel each other and the two Andreev bound states remain at zero energy.

The Josephson current at $T = 0$ is expressed by

$$J = \frac{e\Delta}{2\hbar}\sin\varphi\sum_{k_y}\frac{\Delta|t_n|^2}{\sin\alpha_{nm}}\left[-|\cos X_+| + |\cos X_-|\right]. \tag{6.81}$$

The zero-energy states at the junction interface do not affect the Josephson current at low temperature because the zeros of $\epsilon_\pm$ at the denominator in Eq. (6.76) are removed by the numerator. As a result, the zeros at $\sin\alpha_{nm}$ governs the low-temperature anomaly of the Josephson current. In Fig. 6.9c, we plot the critical value of the

Josephson current as a function of temperature for $z_0 = 5$, where we choose $n = 1$ in the left superconductor. The Josephson current for $m = 1$ increases logarithmically with decreasing Temperature, (Barash and Bobkov 2001, Asano 2002). The low-temperature anomaly becomes weak when the chirality of the two superconductors is opposite to each other as shown with the results for $m = -1$. The results with a chiral f-wave superconductor $m = 3$ show the similar behavior to those with $m = 1$. The low-temperature anomaly is totally absent in the results for $m = -3$.

The Josephson effect between a chiral p-wave superconductor and s-wave superconductor ($\Delta i\hat{\sigma}_2$) shows us a fingerprint of breaking time-reversal symmetry in a chiral p-wave superconductor, (Asano et al. 2004). By taking into account $\Delta_+^* = \Delta(k_x - ik_y)/k_F$ and $\Delta_- = -\Delta(k_x - ik_y)/k_F$ at $\chi = 1$, we find that the Josephson current in such a junction results in

$$J_{s/\text{chiral}-p} = \frac{e}{\hbar} k_B T \sum_{\omega_n>0} \sum_{k_y} \frac{16\bar{k}^2 z_0 \lambda_{\text{SO}} k_y^2}{|\Xi_{\text{SO}}|^2 k_F} \frac{\Delta^2}{(\hbar\omega_n + \Omega)^2} \cos\varphi, \tag{6.82}$$

where we consider spin-orbit coupling in Eq. (6.61) at the junction interface. The Josephson current is proportional to $\cos\varphi$, which reflects the breakdown of time-reversal symmetry in a chiral p-wave superconductor. The presence of the first-order term suggests that $\boldsymbol{d}$ is in perpendicular direction to the two-dimensional plane.

The order parameter of a helical superconductor in two dimensions can be represented as

$$\hat{\Delta}_{\boldsymbol{k}} = \frac{\Delta}{k_F} \begin{bmatrix} -(k_x - i\chi k_y) & 0 \\ 0 & k_x + i\chi k_y \end{bmatrix}, \tag{6.83}$$

where $\chi = \pm 1$ is referred to as "helicity". Thus, it would be possible to say that a Cooper pair with spin-$\uparrow\uparrow$ belongs to the chirality (χ) and that with spin-$\downarrow\downarrow$ belongs to the chirality ($-\chi$). As a result of the coexistence of the two chiral states, helical superconductivity preserves time-reversal symmetry. The tunnel spectra of an NS junction consisting of a helical superconductor are identical to the results displayed in Fig. 6.8a. The Josephson effect between two helical superconductors indicates similar property to that between two chiral superconductors. Thus, it is impossible to distinguish between a helical superconductor and a chiral superconductor by measuring the tunnel spectra and the Josephson current. The Josephson coupling to a spin-singlet s-wave may tell us symmetry information of a spin-triplet superconductor. The Andreev reflection coefficients of a helical superconductor are $\hat{r}_{\text{he}}(R) \propto i(k_x\hat{\sigma}_1 + k_y\hat{\sigma}_2)\hat{\sigma}_2$. Together with the transmission coefficients in Eqs. (6.64)–(6.63) and the Andreev reflection coefficients in Eq. (6.49), we find that the first term in the current formula in Eq. (6.47) vanishes (Fig. 6.10)

$$J_{s/\text{helical}-p} = 0. \tag{6.84}$$

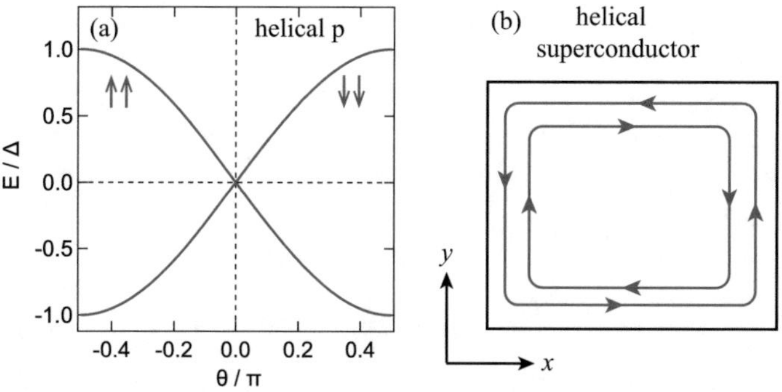

Fig. 6.10 **a** The dispersion of the bound states at a surface of a helical p-wave superconductor. **b** Schematic illustration of the helical edge current. A spin-↓↓ (-↑↑) pair carries the current clockwise (counterclockwise). As a result, the electric current vanishes and the spin current flows

Thus, the lowest order coupling is absent and the current-phase relationship is $J = -J_2 \sin(2\varphi)$ in the case of a helical superconductor. The last property enables us to distinguish a helical superconductor from a chiral one.

References

Asano, Y.: Phys. Rev. B **64**, 224515 (2001). https://doi.org/10.1103/PhysRevB.64.224515
Asano, Y.: J. Phys. Soc. Jpn. **71**(3), 905 (2002). https://doi.org/10.1143/JPSJ.71.905
Asano, Y.: Phys. Rev. B **72**, 092508 (2005). https://doi.org/10.1103/PhysRevB.72.092508
Asano, Y.: Phys. Rev. B **74**, 220501 (2006). https://doi.org/10.1103/PhysRevB.74.220501
Asano, Y., Tanaka, Y., Kashiwaya, S.: Phys. Rev. B **69**, 134501 (2004). https://doi.org/10.1103/PhysRevB.69.134501
Barash, Y.S., Bobkov, A.M., Fogelström, M.: Phys. Rev. B **64**, 214503 (2001). https://doi.org/10.1103/PhysRevB.64.214503
Brydon, P.M.R., Chen, W., Asano, Y., Manske, D.: Phys. Rev. B **88**, 054509 (2013). https://doi.org/10.1103/PhysRevB.88.054509
Maeno, Y., Hashimoto, H., Yoshida, K., Nishizaki, K., Fujita, T., Bednorz, J.G., Lichtenberg, F.: Nature **372**, 532 (1994). https://doi.org/10.1038/372532a0
Millis, A., Rainer, D., Sauls, J.A.: Phys. Rev. B **38**, 4504 (1988). https://doi.org/10.1103/PhysRevB.38.4504
Rice, T.M., Sigrist, M.: J. Phys.: Condens. Matter **7**(47), L643 (1995). https://doi.org/10.1088/0953-8984/7/47/002
Tanaka, Y.: Phys. Rev. Lett. **72**, 3871 (1994). https://doi.org/10.1103/PhysRevLett.72.3871
Tanaka, Y., Kashiwaya, S.: Phys. Rev. B **56**, 892 (1997). https://doi.org/10.1103/PhysRevB.56.892
Volovik, G.E.: J. Exp. Theor. Phys. Lett. **66**(7), 522 (1997). https://doi.org/10.1134/1.567563
Volovik, G.E.: The Universe in a Helium Droplet. Clarendon Press (2003)

Chapter 7
Proximity Effect in a Normal Metal

Abstract When a superconductor is attached to a normal metal, a Cooper pair penetrates into the normal metal. As a result, the normal metal possesses superconducting-like properties such as screening of magnetic fields and decreasing its electric resistance. Such phenomenon is called proximity effect and has been a central issue in the physics of superconductivity. In order to describe the proximity effect correctly, we need to use Green's function technique. In this chapter, however, we will try to explain the essence of the proximity effect phenomenologically by applying the physical picture of the Andreev reflection.

7.1 A Cooper Pair in a Clean Metal

In Fig. 7.1a, we illustrate the classical trajectories of a quasiparticle departing from $\boldsymbol{r}_0$ in a normal metal and reaching at $\boldsymbol{r}_j$ at the NS interface. We assume that the metal is free from any potential disorder and is in the ballistic transport regime. The pair potential belongs to spin-singlet d_{xy}-wave symmetry. The penetration of a Cooper pair into a metal is described by the Andreev reflection. The Andreev reflection coefficient in Eq. (5.34) is summarized as

$$\hat{r}_{\rm he}(\theta) = \frac{|t_n|^2\Gamma_+ e^{-i\varphi}}{1-|r_n|^2\Gamma_+\Gamma_-}\hat{\sigma}_S^\dagger, \quad \Gamma_\pm \equiv \frac{v_\pm}{u_\pm}s_\pm = \frac{-i\Delta_\pm}{\hbar\omega_n + \Omega_{n,\pm}}, \tag{7.1}$$

$$\Delta_+ = \Delta(\theta), \quad \Delta_- = \Delta(\pi-\theta), \tag{7.2}$$

where $k_y = k_F \sin\theta$ and we apply the analytic continuation $E \to i\hbar\omega_n$. The retroreflectivity of a quasiparticle plays an important role in the proximity effect. Let us count the total phase shift of the wave function along the retroreflective trajectories in Fig. 7.1. The wave function at $\boldsymbol{r}_0$ is proportional to $e^{i\boldsymbol{k}\cdot\boldsymbol{r}_0}$. On the way to a place of $\boldsymbol{r}_1$ at the NS interface, an electron gains the phase $e^{i\phi_e}$ with

$$\phi_e = \int_{\boldsymbol{r}_0}^{\boldsymbol{r}_1} d\boldsymbol{l}\cdot\boldsymbol{k}, \tag{7.3}$$

Y. Asano, *Andreev Reflection in Superconducting Junctions*,
SpringerBriefs in Physics, https://doi.org/10.1007/978-981-16-4165-7_7

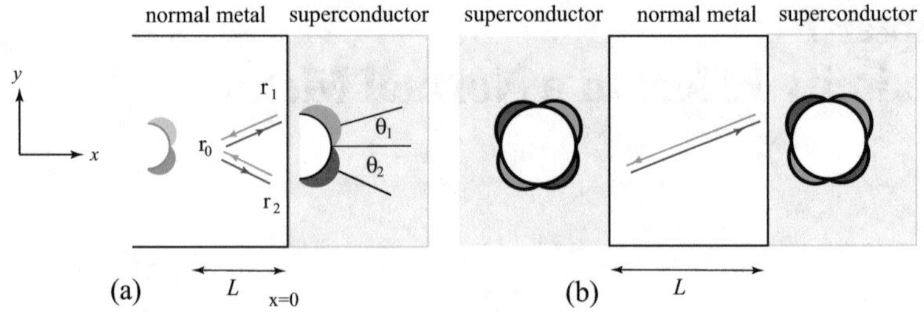

Fig. 7.1 Ballistic motion of a quasiparticle near an NS interface in (**a**). Two typical trajectories are illustrated, where θ_j is the incident angle of a quasiparticle to the NS interface at $\boldsymbol{r}_j$. The angle is measured from the x-axis. In **b**, the second superconductor is attached to the normal metal to make an SNS junction

where $\boldsymbol{l}$ represents a ballistic trajectory between the two points as shown in Fig. 7.1a. At the interface, the Andreev reflection coefficient at $\boldsymbol{r}_1$ would be given by $\hat{r}_{\text{he}}(\theta_1)$ in Eq. (7.1). Finally, on the way back from $\boldsymbol{r}_1$ to $\boldsymbol{r}_0$, a hole gains the phase $e^{i\phi_h}$, with

$$\phi_h = \int_{\boldsymbol{r}_1}^{\boldsymbol{r}_0} d\boldsymbol{l} \cdot \boldsymbol{k} = -\phi_e. \tag{7.4}$$

In the Andreev reflection, the wavenumber remains unchanged and the velocity flips its direction. After traveling along the retroreflective trajectory, a quasiparticle gains the phase factor,

$$p(\boldsymbol{r}_0, \theta_1) = e^{i\phi_h}\, e^{-i\varphi}\, \Gamma_+(\theta_1)\, e^{i\phi_e} = e^{-i\varphi}\, \Gamma_+(\theta_1), \tag{7.5}$$

where we have assumed that the NS interface is transparent perfectly, (i.e., $|t_n| = 1$). The results depend sensitively on the direction of travel of a quasiparticle. By propagating along the retroreflective trajectory between $\boldsymbol{r}_0$ and $\boldsymbol{r}_2$, the phase shift of a quasiparticle becomes

$$p(\boldsymbol{r}_0, \theta_2) = e^{-i\varphi}\Gamma_+(\theta_2). \tag{7.6}$$

The phase factor at $\boldsymbol{r}_0$ depends on the direction of the quasiparticle's motion θ and is proportional to $\Delta_+(\theta)$. In other words, the Andreev reflection copies the d_{xy}-wave pair potential in a superconductor to the place of $\boldsymbol{r}_0$ in a normal metal. The pair potential in real space,

$$\Delta_{\alpha,\beta}(\boldsymbol{r}_1, \boldsymbol{r}_2)\, e^{i\varphi} = g(\boldsymbol{r}_1 - \boldsymbol{r}_2)\, f_{\alpha,\beta}(\boldsymbol{r}_1 - \boldsymbol{r}_2), \quad f_{\alpha,\beta}(\boldsymbol{r}_1, \boldsymbol{r}_2) = \langle \psi_\alpha(\boldsymbol{r}_1)\psi_\beta(\boldsymbol{r}_2)\rangle, \tag{7.7}$$

defines the relation among the pair potential Δ, an attractive interaction g, and the pairing correlation f. Since the attractive interactions are absent in a normal metal, the pair potential is zero. However, the pairing correlation

$$f_{\alpha,\beta}(\boldsymbol{r}_1, \boldsymbol{r}_2) = f(\boldsymbol{r}, \boldsymbol{\rho}) = \frac{1}{V_{\text{vol}}} \sum_{\boldsymbol{k}} f(\boldsymbol{r}, \boldsymbol{k})\, e^{i\boldsymbol{k}\cdot\boldsymbol{\rho}}, \tag{7.8}$$

with $\boldsymbol{r} = (\boldsymbol{r}_1 + \boldsymbol{r}_2)/2$ and $\boldsymbol{\rho} = \boldsymbol{r}_1 - \boldsymbol{r}_2$, describes the existence of a Cooper pair in the metal. The phase factor $p(\boldsymbol{r}, \theta)$ is a part of the correlation function $f(\boldsymbol{r}, \boldsymbol{k})$. The results in Eqs. (7.5) and (7.6) suggest that the pairing correlation at $\boldsymbol{r}_0$ belongs to a d_{xy}-wave symmetry. Namely, a Cooper pair of spin-singlet d-wave symmetry penetrates into a ballistic normal metal as schematically illustrated in Fig. 7.1a.

Since the pair potential is absent in a normal metal, the phase coherence of a quasiparticle supports a Cooper pair. At a finite temperature, however, the thermal broadening of the Fermi level destroys the phase coherence. When the energy of an electron is distributed in a range of $\hbar\omega_n = (2n+1)\pi k_B T$ around ϵ_F, the relation

$$\frac{\hbar^2(k_F + \delta k)^2}{2m} \approx \epsilon_F + \hbar\omega_n \tag{7.9}$$

represents how the thermal broadening modifies the wavenumber. We obtain $\delta k = \omega_n/v_F$. When an electron comes into a normal metal from $x = 0$ and propagates to $x = L$, the wave function of a quasiparticle just at the Fermi level is given by $\exp\{ik_F L\}$. The wave function of the quasiparticle at an energy $\epsilon_F + \hbar\omega_n$ can be described by $\exp\{i(k_F + \delta k)L\}$. A relation $\delta k L < 1$ must be satisfied so that the two partial waves of a quasiparticle interfere with each other. This argument at the lowest Matsubara frequency gives a length scale

$$\xi_T^C = \frac{\hbar v_F}{2\pi k_B T}, \tag{7.10}$$

which represents the spatial range of the phase coherence of a quasiparticle in a clean metal.

When an another superconductor is attached to the normal metal as shown in Fig. 7.1b, a Cooper pair penetrates from the left superconductor, propagates a normal metal, and transmits to the right superconductor. This process explains the Josephson effect in a clean superconductor/normal metal/superconductor (SNS) junction. The Andreev reflection coefficients in the lowest order process in Fig. 4.7b can be calculated as

$$\begin{aligned} A_1 =& \frac{ie}{\hbar} \sum_{k_y} k_B T \sum_{\omega_n} \mathrm{Tr}\left[\Gamma_+ e^{-i\varphi_R} \cdot t_N e^{-\omega_n L/\hbar v_F} \cdot \Gamma_+ e^{i\varphi_L} \cdot t_N e^{-\omega_n L/\hbar v_F} - \text{c.c.}\right], \\ =& \frac{4e}{\hbar} \sum_{k_y} k_B T \sum_{\omega_n} \sin(\varphi)(-1) e^{-2\omega_n L/\hbar v_F} \left(\frac{\Delta_+}{\omega_n + \Omega_+}\right)^2, \end{aligned} \tag{7.11}$$

where $t_N = 1$ is the transmission coefficient at the normal metal. In a clean SNS junction, the higher order processes such as in Fig. 4.7c also contribute to the Josephson current. It is possible to describe the total Josephson current as

$$J = \frac{4e}{\hbar} \sum_{k_y} k_B T \sum_{\omega_n} \sum_{m=1}^{\infty} \sin(m\varphi)(-1)^m e^{-2\omega_n mL/\hbar v_F} \left(\frac{\Delta_+}{\omega_n + \Omega_+} \right)^{2m}. \quad (7.12)$$

Since Δ_+^2 enters the current expression, the sign change of the pair potential does not play any important role in Eq. (7.12). Thus, we replace Δ_+ by Δ approximately. The summation over the Matsubara frequency at $T = 0$ can be replaced by the integration as

$$J = \frac{4e}{\hbar} N_c \sum_{m=1}^{\infty} \sin(m\varphi)(-1)^m \frac{1}{\pi} \int_0^{\infty} d\omega \, e^{-2\omega mL/\hbar v_F} \left(\frac{\sqrt{\omega^2 + \Delta^2} - \omega}{\Delta} \right)^{2m}, \quad (7.13)$$

with $\sum_{k_y} = N_c$. The integrand is a product of two functions which have a peak at $\omega = 0$. The decay energy of the exponential function is $\hbar v_F / L$ and that of power function is Δ. When $\hbar v_F / L \gg \Delta$, the integration can be carried out as

$$J = \frac{e\,\Delta}{\hbar} N_c \sum_{m=1}^{\infty} \sin(m\varphi) \frac{(-1)^{m+1}}{\pi} \frac{8m}{(2m+1)(2m-1)} = \frac{e\,\Delta}{\hbar} N_c \sin\left[\frac{\varphi}{2}\right] \quad (7.14)$$

because of Eq. (4.80). The results were derived by Kulik and Omel'yanchuk (1977). The condition $\hbar v_F / L \gg \Delta$ is identical to $L \ll \xi_0$ with $\xi_0 = \hbar v_F / \pi\Delta \simeq \hbar v_F / 2\pi k_B T_c$ being the coherence length. On the other hand, in the long junction limit $L \gg \xi_0$, we obtain

$$J = \frac{e\,v_F}{\pi L} N_c \sum_{m=1}^{\infty} \sin(m\varphi) \frac{2(-1)^{m+1}}{m} = \frac{e\,v_F}{L} N_c \frac{\varphi}{\pi}. \quad (7.15)$$

Such linear current-phase relationship was first discussed by Ishii (1970). In a SNS junction in the ballistic regime, Eqs. (7.14) and (7.15) represent the Josephson current between two identical superconductors at $T = 0$. The results are insensitive to the symmetry of pair potential.

7.2 Andreev Reflection and Diffusive Motion

The random impurities in a normal metal affect drastically the low-energy transport properties in an NS junction consisting of an unconventional superconductor. Let us begin with a schematic illustration of the Andreev reflection into a dirty normal metal. The motion of a quasiparticle in a random media is described by the diffusion equation

$$\left(D\nabla^2 - \partial_t\right) F(\boldsymbol{r}, t) = 0, \quad F(\boldsymbol{r}, 0) = F_0\, \delta(\boldsymbol{r}). \tag{7.16}$$

The diffusion constant D is given by

$$D = \frac{v_F^2 \tau}{d} = \frac{v_F \ell}{d}, \quad \ell = v_F\, \tau, \tag{7.17}$$

where $d = 1 - 3$ denotes the spatial dimension, $v_F = \hbar k_F/m$ is the Fermi velocity, and τ is the mean free time due to the elastic scattering by random impurities. The mean free path defined by $\ell = v_F \tau$ characterizes the smallest length scale of the diffusion equation. The solution of the diffusion equation

$$F(\boldsymbol{r}, t) = \frac{F_0}{(4\pi D t)^{d/2}} \exp\left(-\frac{|\boldsymbol{r}|^2}{4Dt}\right) \tag{7.18}$$

tells us characteristic features of a quasiparticle's motion. The diffusive motion is isotropic in real space. Although the velocity of a quasiparticle is v_F, it takes L^2/D (sec) to propagate over a distance of L. Therefore, the total length of such a diffusive trajectory is $L' = (L^2/D)v_F$ while an electron propagates to diffuse by L. The thermal broadening of the Fermi distribution function is a source of the decoherence. As we discussed below in Eq. (7.9), $\delta k L' < 1$ is necessary for a quasiparticle to keep its phase coherence. Therefore,

$$\xi_T^D = \sqrt{\frac{\hbar D}{2\pi k_B T}} \tag{7.19}$$

represents the length scale for phase coherent phenomena in a diffusive metal and is called thermal coherence length. Energetically, $E_{\mathrm{Th}} = \hbar D/L^2$ is called the Thouless energy which represents the critical energy scale for the phase coherent phenomena.

As shown in Fig. 4.4, the retroreflectivity of a quasiparticle is a key feature of the proximity effect. In Fig. 7.2a, we illustrate two classical retroreflective trajectories of a quasiparticle departing from a point $\boldsymbol{r}_0$ and reaching the NS interface at $\boldsymbol{r}_j$. We count the phase shift of a quasiparticle moving along a retroreflective trajectory. On the way to the NS interface, an electron gains the phase of $e^{i\phi_e}$ with

$$\phi_e = \int_{\boldsymbol{r}_0}^{\boldsymbol{r}_1} d\boldsymbol{l} \cdot \boldsymbol{k} = \int_{\boldsymbol{r}_0}^{\boldsymbol{R}_1} d\boldsymbol{l}_1 \cdot \boldsymbol{k}_1 + \int_{\boldsymbol{R}_1}^{\boldsymbol{R}_2} d\boldsymbol{l}_2 \cdot \boldsymbol{k}_2 + \cdots + \int_{\boldsymbol{R}_N}^{\boldsymbol{r}_1} d\boldsymbol{l}_N \cdot \boldsymbol{k}_N, \tag{7.20}$$

where $\boldsymbol{l}_j$ represents a classical trajectory between two impurities at $\boldsymbol{R}_{j-1}$ and $\boldsymbol{R}_j$ as shown in Fig. 7.2a. The direction of momentum $\boldsymbol{k}_j$ changes at every scattering event by an impurity. At the interface, the Andreev reflection coefficient at $\boldsymbol{r}_1$ would be given by $\hat{r}_{\mathrm{he}}(\theta_1)$ in Eq. (7.1). Finally, on the way back from $\boldsymbol{r}_1$ to $\boldsymbol{r}_0$, a hole gains the phase of $e^{i\phi_h}$, with

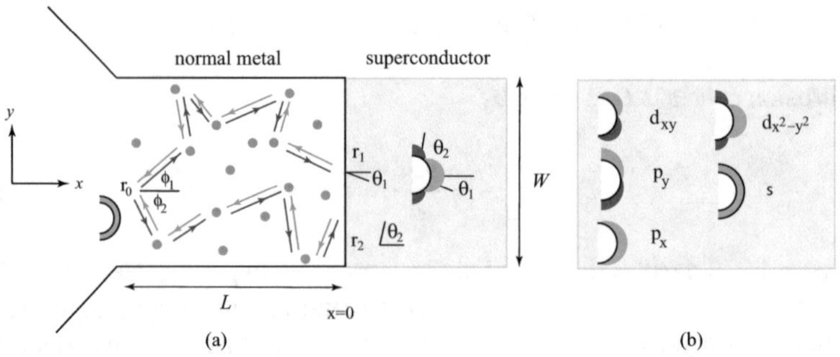

Fig. 7.2 **a** Diffusive motion of a quasiparticle near an NS interface. Two typical trajectories are illustrated, where θ_j is the incident angle of a quasiparticle into a superconductor at $\boldsymbol{r}_j$. The pair potential is shown only for $k_x > 0$ in (**a**) and **b** because such part is copied to a normal metal by the Andreev reflection

$$\phi_h = \int_{\boldsymbol{r}_1}^{\boldsymbol{R}_N} d\boldsymbol{l}_N \cdot \boldsymbol{k}_N + \cdots + \int_{\boldsymbol{R}_2}^{\boldsymbol{R}_1} d\boldsymbol{l}_2 \cdot \boldsymbol{k}_2 + \int_{\boldsymbol{R}_1}^{\boldsymbol{r}_0} d\boldsymbol{l}_1 \cdot \boldsymbol{k}_1 = -\phi_e. \tag{7.21}$$

The two phases ϕ_e and ϕ_h cancel to each other even in the diffusive metal. After traveling along a retroreflective trajectory, a quasiparticle wave function gains a phase factor,

$$p_1(\boldsymbol{r}_0) = \hat{r}_{\text{he}}(k_{y_1})\, e^{-2L/\xi_T^D}, \tag{7.22}$$

where $k_{y_1} = k_F \sin\theta_1$. The results depend on the angle of incidence θ_1 to the NS interface at $\boldsymbol{r}_1$, but do not depend on the traveling angle ϕ_1 at $\boldsymbol{r}_0$. As shown in Fig. 7.2a, another part of such a quasiparticle travels along another retroreflective trajectory which departs from $\boldsymbol{r}_0$ and reaches at $\boldsymbol{r}_2$ on the NS interface. Such partial wave function gains a phase factor

$$p_2(\boldsymbol{r}_0) = \hat{r}_{\text{he}}(k_{y_2})\, e^{-2L/\xi_T^D}, \tag{7.23}$$

with $k_{y_2} = k_F \sin\theta_2$. The phase shift at $\boldsymbol{r}_0$ should be averaged over all possible diffusive trajectories as

$$\langle p(\boldsymbol{r}_0)\rangle = \frac{1}{N_c}\sum_{k_y} \hat{r}_{\text{he}}(k_y)\, e^{-2L/\xi_T^D} = \frac{1}{2}\int_{-\pi/2}^{\pi/2} d\theta\, \cos\theta\, \hat{r}_{\text{he}}(\theta)\, e^{-2L/\xi_T^D}. \tag{7.24}$$

The Andreev reflection copies the pair potential averaged over θ into a dirty normal metal. As a result of averaging, the phase shift in Eq. (7.24) at $\boldsymbol{r}_0$ is isotropic in momentum space independent of the initial direction of ϕ in Fig. 7.2a. This consideration implies that the pairing correlation at $\boldsymbol{r}_0$ belongs to s-wave symmetry irrespective of the symmetries of pair potential.

Even so, the properties of the pairing correlation at $\boldsymbol{r}_0$ depends sensitively on the symmetry of pair potential in the superconductor. Here, we consider the limit of the low frequency $\hbar\omega_n \ll \Delta$ at zero temperature. The coherence factor becomes $\Gamma_+ \to -i\, e^{-i\varphi} s_+$ with $s_+ = \mathrm{sgn}[\Delta_+(\theta)]$. We also assume that the transparency of the NS interface is small, ($|t_n| \ll 1$). In an NS junction of a s-wave superconductor, we obtain

$$\langle p_s(\boldsymbol{r}_0\rangle = -i\, e^{-i\varphi}\, \frac{1}{2}\, |t_n|^2\, i\hat{\sigma}_2. \tag{7.25}$$

The Andreev reflection copies also the spin configuration of a Cooper pair and the phase in the superconductor to the dirty metal as shown in Eq. (7.1). The result suggests that a spin-singlet s-wave Cooper pair penetrates into the metal. In a spin-singlet $d_{x^2-y^2}$-wave case as shown in Fig. 7.2a, the phase factor at $\boldsymbol{r}_0$ becomes

$$\langle p_{d_{x^2-y^2}}(\boldsymbol{r}_0)\rangle = -i\, e^{-i\varphi}\, \left(\sqrt{2}-1\right)\, |t_n|^2 i\hat{\sigma}_2. \tag{7.26}$$

The amplitude of Eq. (7.26) is slightly smaller than that of Eq. (7.25) because the pair potential changes its sign on the Fermi surface. The impurity scatterings make physics isotropic in real space. In other words, impurities destroy such anisotropic Cooper pairs as p-wave and d-wave symmetry. In a d_{xy}-symmetry, we find

$$\langle p_{d_{xy}}(\boldsymbol{r}_0)\rangle = -i\, e^{-i\varphi}\, i\, \hat{\sigma}_2\, \frac{1}{2} \int_{-\pi/2}^{\pi/2} d\theta\, \cos\theta\, \mathrm{sgn}\,[\sin 2\theta] = 0, \tag{7.27}$$

because the pair potential $\Delta \sin 2\theta$ is an odd function of θ. Namely, the proximity effect is absent in a metal attached to a d_{xy}-wave superconductor, (Asano 2001). In the same manner, the proximity effect is absent in a metal attached to a p_y-wave superconductor because

$$\langle p_{p_y}(\boldsymbol{r}_0)\rangle = -i\, e^{-i\varphi}\, \hat{\sigma}_S\, \frac{1}{4} |t_n|^2 \int_{-\pi/2}^{\pi/2} d\theta\, \cos\theta\, \mathrm{sgn}\,[\sin\theta] = 0, \tag{7.28}$$

Thus, the presence or absence of the proximity effect depends on an orientation angle of the pair potential to the NS interface. Finally, in spin-triplet p_x-wave symmetry, we find

$$\langle p_{p_x}(\boldsymbol{r}_0)\rangle = -i\, e^{-i\varphi} \hat{\sigma}_S. \tag{7.29}$$

The amplitude of the phase factor is much larger than that of a s-wave junction because of the resonant transmission through the Andreev bound states at the NS interface. The orbital part is even-parity s-wave symmetry and the spin part is spin-triplet symmetry class. At the first glance, such a Cooper pair does not meet the requirement of the Fermi-Dirac statistics of electrons because both the spin and orbital parts of the pairing correlation are even under the permutation of two electrons. It was pointed out that such pairing correlation function can be antisymmetric under the permutation of two "times" of electrons, (Berezinskii 1974). In the Matsubara representation, the pairing function of such a Cooper pair is an odd function of ω_n. This argument indicates that an odd-frequency spin-triplet even-parity s-wave pair penetrates into a dirty metal. Unfortunately, Green's function technique is necessary to discuss the frequency symmetry of a Cooper pair. Instead of going into theoretical details, we explain the anomalous transport property due to an odd-frequency Cooper pair.

7.3 Anomalous Proximity Effect

In Chap. 5, we discussed the formation of the Andreev bound states at a surface of an unconventional superconductor. In Sect. 7.2, we discussed the presence or the absence of the proximity effect in a dirty normal metal attached to an unconventional superconductor. These phenomena depend sensitively on how the pair potential changes sign under inversion of junction geometries. In two-dimensional junctions in Fig. 5.1, the relation

$$\Delta(-k_x, k_y) = -\Delta(k_x, k_y) \tag{7.30}$$

represents a condition for the appearance of the Andreev bound states at a surface parallel to the y-direction. Another relation

$$\Delta(k_x, -k_y) = -\Delta(k_x, k_y) \tag{7.31}$$

describes the absence of the proximity effect in a dirty metal attached to the superconductor at $x = 0$. Table 7.1 shows the relation between the presence or absence of these coherent phenomena and the symmetry of a superconductor. The surface Andreev bound states are possible in a d_{xy}-wave superconductor and a p_x-wave superconductor. The proximity effect is present in an s-wave junction and a p_x-wave junction. The two coherent phenomena occur simultaneously in a p_x-wave junction. In this section, we will show that a p_x-wave junction exhibits unusual low-energy transport properties as a result of the interplay between the formation of Andreev bound states and the penetration of a Cooper pair into a normal metal.

In Fig. 7.3a, we plot the total resistance of an NS junction R_{NS} as a function of the normal resistance of a metal R_{N}, where $R_0 = (2e^2 N_c/h)^{-1}$ is the Sharvin resistance of the junction. The resistance is calculated by using quasiclassical Green's function method (Kopnin 2001). Theoretical details are given in Tanaka and Kashiwaya (2004). In classical mechanics, two resistances contribute to R_{NS} independently as

$$R_{\text{NS}} = R_{\text{N}} + R_{\text{B}}, \tag{7.32}$$

Table 7.1 The relation between the symmetry of pair potential and two types of phase coherent phenomena in an NS junction consisting of a superconductor $x > 0$ and a dirty normal metal $x < 0$ in Fig. 7.2a. At the first row, "○" means the presence of zero-energy Andreev bound states at a surface of a superconductor. The pair potential must be an odd function of k_x to host the Andreev bound states. At the second row, "○" means the penetration of a Cooper pair into a dirty metal. When the pair potential has an even function part of k_y, the proximity effect occurs in the dirty metal. A $d_{x^2-y^2}$-wave pair potential belongs to the same class as an s-wave pair potential

	$s, d_{x^2-y^2}$	d_{xy}	p_y	p_x
Surface ABS	–	○	–	○
Proximity effect	○	–	–	○

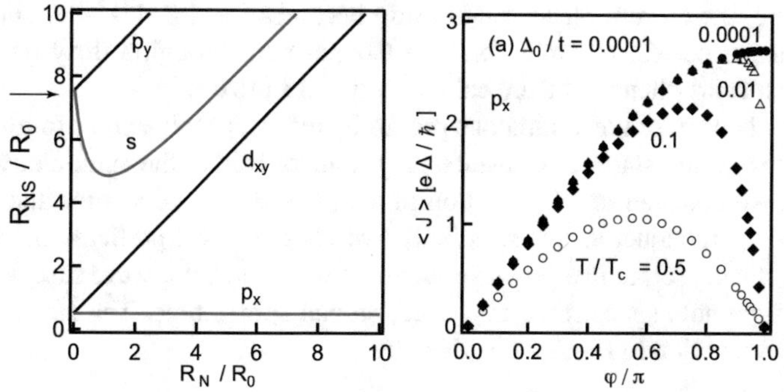

(a) Total resistance of a NS junction

(b) Current-phase relationship in a p_x-wave SNS junction

Fig. 7.3 **a** The total resistance of a dirty NS junction is plotted as a function of the resistance in a normal metal, where $R_0 = (2e^2/h)^{-1}$ is the Sharvin resistance of the junction. We attach a lead wire for $x < -L$ to measure the conductance of a junction in Fig. 7.2. **b** The current-phase relationship in a dirty SNS junction for a p_x-wave symmetry. We attach another p_x-wave superconductor to the normal metal at $x = -L$ in Fig. 7.2

where the R_N is resistance in the normal metal and $R_B \simeq 7.56R_0$ indicated by an arrow is the resistance of the potential barrier at the NS interface. The results for a p_y-wave junction increases linearly with R_N. In a p_y-wave NS junction, both the proximity effect in a metal and the Andreev bound states at the surface are absent. Therefore, the total resistance agrees with the classical relationship in Eq. (7.32). The deviation of R_{NS} from Eq. (7.32) would be expected in the presence of phase coherent effects. In a d_{xy}-wave case, the results show that $R_{NS} = R_N + R_0/2$ holds instead of Eq. (7.32). Although the proximity effect is absent in a metal, the resonant transmission through the Andreev bound states reduces the resistance at the interface to $R_0/2$. In a s-wave junction, R_{NS} first decreases from R_B with the increase of R_N and then increases. Such reentrant behavior is a result of the usual proximity effect in a metal. Impurities near the interface enhance the Andreev reflection probability, which explains the suppression of R_{NS} for $R_N < R_B$. In this range, the dirtier junction has the smaller resistance. Finally, for a p_x-wave symmetry, $R_{\rm NS}$ is constant at $R_0/2$ independently of $R_{\rm N}$. The differential conductance at zero bias is given by

$$G_{\rm NS} = R_{\rm NS}^{-1} = \frac{4e^2}{h} N_c, \tag{7.33}$$

where N_c is the number of propagating channels per spin. The results imply as if a normal metal loses its resistance due to the the interplay between the two phase coherent effects. The low-temperature anomaly would be expected also in a dirty SNS junction consisting of two p_x-wave superconductors. The results in Fig. 7.3b show the Josephson current in such an SNS junction plotted as a function of the phase difference between two superconductors, (Asano et al. 2006). At a very low temperature, the current-phase relationship becomes fractional $J = J_c \sin(\varphi/2)$ as a result of the perfect transmission of a Cooper pair through a dirty metal. Such unusual transport phenomena are called anomalous proximity effect.

The mechanism of the anomalous proximity effect is explained as follows. Since the Andreev bound states at a surface play a major role in the anomalous proximity effect, we consider an NS junction of a p_x-wave superconductor and that of a d_{xy}-wave superconductor. For simplicity, we choose 2×2 particle-hole space by extracting $\uparrow$-spin sector in a p_x-wave superconductor. In a d_{xy}-wave case, we choose 2×2 particle-hole space for spin-$\uparrow$ electron and spin-$\downarrow$ hole. The Hamiltonian of such an NS junction in Fig. 7.2a is given by

$$\hat{H} = \begin{bmatrix} h_{\boldsymbol{r}} & \Delta(\boldsymbol{r}) \\ \Delta(\boldsymbol{r}) & -h_{\boldsymbol{r}} \end{bmatrix} = h_{\boldsymbol{r}}\, \hat{\tau}_3 + \Delta(\boldsymbol{r})\hat{\tau}_1, \tag{7.34}$$

$$\Delta(\boldsymbol{r}) = \begin{cases} \Delta\, \Theta(x)\, \frac{-i}{k_F}\frac{\partial}{\partial x} & : p_x\text{−wave} \\ \Delta\, \Theta(x)\, \frac{-1}{k_F^2}\frac{\partial^2}{\partial x \partial_y} & : d_{xy}\text{−wave}, \end{cases} \tag{7.35}$$

$$h_{\boldsymbol{r}} = \xi_{\boldsymbol{r}} + v_0\delta(x) + V_{\rm imp}(\boldsymbol{r})\Theta(-x), \tag{7.36}$$

$$\xi_{\boldsymbol{r}} = -\frac{\hbar^2\nabla^2}{2m} - \epsilon_F, \quad V_{\rm imp}(\boldsymbol{r}) = \sum_{\boldsymbol{r}_i} v_{\rm imp}\delta(\boldsymbol{r} - \boldsymbol{r}_i), \tag{7.37}$$

where $\hat{\tau}_j$ for $j = 1 - 3$ is Pauli's matrix in particle-hole space. Chiral symmetry of the Hamiltonian is described by

$$\left\{\hat{H}, \hat{\Lambda}\right\} = 0, \quad \hat{\Lambda} = \hat{\tau}_2. \tag{7.38}$$

Since $\hat{\Lambda}^2 = \hat{1}$, the eigenvalue of $\hat{\Lambda}$ is either $\lambda = 1$ or $\lambda = -1$. In what follows, we refer λ as chirality. The eigenstate vectors are

$$\frac{1}{\sqrt{2}}\begin{bmatrix} 1 \\ -i \end{bmatrix}, \quad \frac{1}{\sqrt{2}}\begin{bmatrix} 1 \\ i \end{bmatrix}, \tag{7.39}$$

for $\lambda = -1$ and $\lambda = 1$, respectively. Here, we summarize two important features of the eigenstates of $\hat{H}$ preserving chiral symmetry, (Sato et al. 2011).

(i) The eigenstates of $\hat{H}$ at zero energy are simultaneously the eigenstates of $\hat{\Lambda}$. Namely, the eigenvectors at zero energy $\psi_{\nu_0,\lambda}(\boldsymbol{r})$ satisfy

$$\hat{H}\,\psi_{\nu_0,\lambda}(\boldsymbol{r}) = 0, \quad \hat{\Lambda}\,\psi_{\nu_0,\lambda}(\boldsymbol{r}) = \lambda\,\psi_{\nu_0,\lambda}(\boldsymbol{r}), \tag{7.40}$$

where ν_0 is the label of zero-energy states.

(ii) In contrast to the zero-energy states, the nonzero-energy states are not the eigenstates of $\hat{\Lambda}$. They are described by the linear combination of two states: one has $\lambda = 1$ and the other has$\lambda = -1$ as

$$\psi_{E\neq 0} = a_- \begin{bmatrix} 1 \\ -i \end{bmatrix} + a_+ \begin{bmatrix} 1 \\ i \end{bmatrix}. \tag{7.41}$$

In addition, the relation $|a_+| = |a_-|$ is always satisfied. A positive chiral zero-energy state and a negative chiral zero-energy state couple one-by-one and form two nonzero-energy states.

Below, we will prove the robustness of the highly degenerate zero-energy states in a dirty normal by taking these features into account. We also show that resonant transmission of a Cooper pair via such degenerate zero-energy states causes the perfect Andreev reflection and the quantization of conductance.

By repeating the argument at Eq. (5.54), the wave function of the surface bound states at zero energy is described by

$$\psi_{k_y}(\boldsymbol{r}) = C \begin{bmatrix} 1 \\ -i\,s_+ \end{bmatrix} \sin(k_x x)\, e^{-x/\xi_0}\, e^{ik_y y}, \quad s_+ = \frac{\Delta_{\boldsymbol{k}}}{|\Delta_{\boldsymbol{k}}|}. \tag{7.42}$$

It is easy to confirm property (i) because this wave function is the eigenstate of Λ belonging to $\lambda = -s_+$. In quantum mechanics, the large degree of degeneracy at quantum states is a direct consequence of the high symmetry of Hamiltonian. The degeneracy is lifted by adding perturbations breaking the symmetry. In the present case, both p_x- and d_{xy}-superconductors host N_c-fold degenerate bound states at zero energy as shown in Fig. 5.6b, c. Such highly degenerate states are stable because translational symmetry in the y-direction is preserved in a clean superconductor. However, introducing random impurities at a surface may lift the degeneracy because they break translational symmetry. At a surface of a d_{xy}-wave superconductor, zero-energy states for $k_y > 0$ belong to the negative chirality, whereas those for $k_y < 0$ belong to the positive chirality. According to property (ii), a positive chiral zero-energy state and a negative chiral zero-energy state can be paired by impurity scatterings and form two nonzero-energy states. As a result, all the zero-energy states disappear in the presence of random impurities. On the other hand, in a p_x-wave superconductor, all the zero-energy states belong to the negative chirality. Property (ii) suggests that the degeneracy of such pure chiral states is robust even in the presence of impurities because their chiral partners for forming nonzero-energy states are absent. The number of zero-energy states in the presence of impurities can be described by the absolute value of

$$N_{\mathrm{ZES}} = N_+ - N_-, \tag{7.43}$$

where $N_\pm$ is the number of zero-energy states belonging to $\lambda = \pm 1$. The index N_{ZES} is an invariant as long as the Hamiltonian preserves chiral symmetry. Thus, N_{ZES} calculated in the absence of impurities does not change even in the presence of impurities. In mathematics, the resulting index N_{ZES} is known as an invariant constructed from solutions of a differential equation. Simultaneously, the index is identical to the summation of the winding number over k_y as

$$N_{\mathrm{ZES}} = -\sum_{k_y} \mathcal{W}(k_y), \tag{7.44}$$

where the winding number is defined in Eq. (5.93) and is calculated as Eq. (5.94). Namely, the index is a topological invariant. In mathematics, such an integer number N_{ZES} is called Atiyah-Singer's index. We find $N_{\mathrm{ZES}} = 0$ for a d_{xy}-wave superconductor and $N_{\mathrm{ZES}} = -N_c$ for a p_x-wave superconductor.

The remaining task is to show why the zero-bias conductance in Eq. (7.33) is quantized as

$$G_{\mathrm{NS}} = \frac{4e^2}{h} |N_{\mathrm{ZES}}|, \tag{7.45}$$

in the limit of $R_{\mathrm{N}} \to \infty$, (Ikegaya et al. 2016). In a normal metal, the wave function at $E = 0$ can be exactly represented as

$$\psi_{\mathrm{N}}(\boldsymbol{r}) = \sum_{k_y} \left[\begin{pmatrix} 1 \\ r_{\mathrm{he}}(k_y) \end{pmatrix} e^{ik_x x} + \begin{pmatrix} r_{\mathrm{ee}}(k_y) \\ 0 \end{pmatrix} e^{-ik_x x} \right] e^{ik_y y}, \tag{7.46}$$

where the reflection coefficients are given in Eqs. (5.33) and (5.34). They become

$$r_{\mathrm{ee}} = 0, \quad r_{\mathrm{he}} = -is_+, \tag{7.47}$$

in the limit of $E \to 0$ for a p_x-wave NS junctions. By substituting the coefficients into the wave function, we find that the zero-energy states in a clean metal,

$$\psi_{\mathrm{N}}(\boldsymbol{r}) = \sum_{k_y} \begin{pmatrix} 1 \\ -is_+ \end{pmatrix} e^{ik_x x} e^{ik_y y} = \begin{pmatrix} 1 \\ -i \end{pmatrix} \sum_{k_y} e^{i\boldsymbol{k}\cdot\boldsymbol{r}}, \tag{7.48}$$

belongs to $\lambda = -1$. As a result, all the zero-energy states in the normal metal have the negative chirality. In other words, the Andreev reflection copies the chirality of

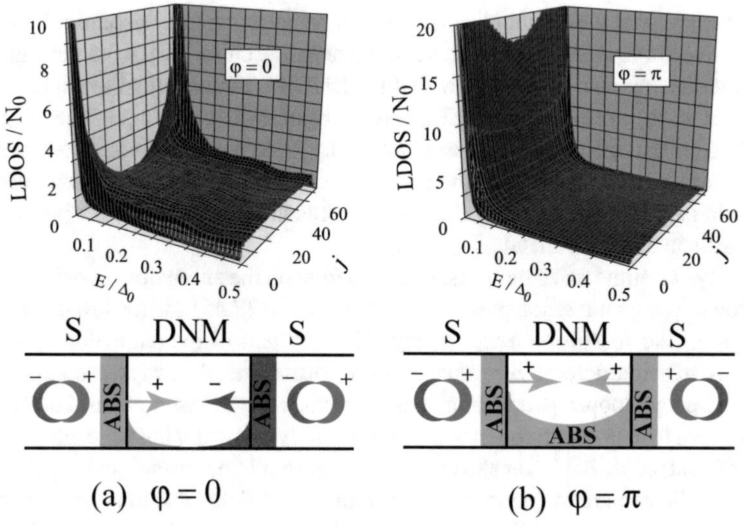

Fig. 7.4 The local density of states (LDOS) at a dirty normal metal (DNM) in a SNS junction of spin-triplet p_x-wave superconductors, where N_0 is the normal density of states per unit area at the Fermi level, Δ_0 is the amplitude of pair potential at zero temperature, the left NS interface is at $j = 0$, and the right NS interface is at $j = 70$. The phase difference across the SNS junction is chosen as $\varphi = 0$ in (**a**) and $\varphi = \pi$ in (**b**). The lower schematic pictures represent the spatial profile of the positive chiral Andreev bound states (ABSs) at zero energy (+) and that of negative chiral ABSs at zero energy (−)

the zero-energy states at a surface of the p_x-wave superconductor to the zero-energy states in the normal metal. This argument is valid even in the presence of impurities because the pure chiral zero-energy states can retain their high degeneracy. It is not easy to represent how the orbital part of the wave function depends on $\boldsymbol{r}$ in a dirty metal. The vector part of the wave function, however, remains unchanged from that in Eq. (7.48). In the vector part, 1 at the first row is the amplitude of the incoming wave into the NS interface and $-i$ at the second row is necessary for the vector to be an eigenstate of $\hat{\Lambda}$. More importantly, the $-i$ is derived from the perfect Andreev reflection, which explains the quantization of the zero-bias conductance in a dirty NS junction. For the perfect Andreev reflection to occur, the wave function at zero energy must be extended throughout the dirty metal. Chiral symmetry of the BdG Hamiltonian is responsible for the high degeneracy at the zero-energy states in a dirty metal. In Fig. 7.4, we show the local density of states (LDOS) in a dirty normal metal (DNM) sandwiched by two p_x-wave superconductors, where $j = 0$ and $j = 70$ indicate the place of left NS interface and that of right NS interface, respectively. The LDOS is calculated by using lattice Green's function method on the two-dimensional tight-binding lattice. The schematic picture at the bottom shows the chiral property of the Andreev bound states. At $\varphi = 0$ in (a), the chirality of ZESs at a surface of the left superconductor is positive, whereas that at a surface of the right superconductor is negative. As schematically illustrated in the lower picture,

ZESs disappear at the center of normal metal because impurity potential hybridizes the positive chiral ZES from the left and the negative chiral ZES from the right. The calculated results in Fig. 7.4a show that the LDOS is almost equal to N_0 at $j = 35$. At $\varphi = \pi$ in (b), the chirality of ZESs on the right is changed to positive. The large peak at $E = 0$ in the LDOS suggests that all the ZESs penetrate into a normal metal, stay at zero energy, and bridge the two superconductors. The Josephson current at $\varphi \approx \pi$ in Fig. 7.3b is a result of the resonant transmission of a Cooper pair through such ZESs in a normal metal.

Finally, we summarize our present knowledge on the anomalous proximity effect as follows. The quantization of the conductance in Eq. (7.45) and the low-temperature anomaly in the Josephson effect in Fig. 7.3 are a part of the anomalous proximity effect. A quasiparticle at zero energy in a normal metal always accompanies an odd-frequency Cooper pair, (Tanaka and Golubov 2007, Asano et al. 2007). It has been known that an odd-frequency pair is thermodynamically unstable and paramagnetic, (Asano et al. 2011, Tanaka et al. 2005). A small unconventional superconductor may indicate paramagnetic response to magnetic fields due to an odd-frequency Cooper pair at its surface (Suzuki et al. 2015). Experimental studies are desired to confirm such theoretical predictions.

7.4 Fluctuations of the Josephson Current

In order to measure the conductance in metals in experiments, we prepare a number of samples of metal wire and measure the conductance of all the samples. The ensemble average of the conductance is defined by

$$\langle G \rangle = \frac{1}{N_\mathrm{s}} \sum_{j=1}^{N_\mathrm{s}} G^{(j)}, \tag{7.49}$$

where $G^{(j)}$ is the conductance measured for the jth sample and N_s is the number of measured samples. The ensemble average of conductance should be

$$\langle G \rangle = G_\mathrm{N} \equiv \frac{S}{L} \sigma_\mathrm{D}, \quad \sigma_\mathrm{D} = 2e^2 N_0 D, \tag{7.50}$$

where σ_D is Drude's conductivity, N_0 is the density of states per unit volume per spin at the Fermi level, and D is the diffusion constant of the diffusive metal. No quantum mechanical considerations are needed to derive Drude's conductivity. In classical mechanics, sample-to-sample fluctuations of conductance

$$\delta G = \sqrt{\langle G^2 \rangle - \langle G \rangle^2} \tag{7.51}$$

is derived from the fluctuations in the sample dimensions. If all the samples have the same length L and the same cross section S exactly, the fluctuations are expected to be zero. However, from a quantum mechanical point of view, the sample-to-sample fluctuations in conductance originate from the sample-specific random arrangement of impurities. An electron memorizes the microscopic impurity configuration of a sample in the phase of its wave function, (Lee and Stone 1985, Al'tshuler 1985, Imry 2002). Therefore, the fluctuations can be observed at a low temperature satisfying $k_B T \ll E_{\mathrm{Th}}$, where

$$E_{\mathrm{Th}} = \frac{\hbar D}{L^2}, \tag{7.52}$$

is the Thouless energy of a conductor. The conductance measured in a sample is slightly different from those in other samples because an electron preserves the phase coherence while it propagates of a metal in such a low temperature. The fluctuations at zero temperature can be described as

$$\delta G = \frac{e^2}{h} C_G, \tag{7.53}$$

where C_G is a constant of the order of unity. This fact is called universal conductance fluctuations because C_G depends only on the spatial dimensionality of a conductor and the symmetry of Hamiltonian, and not on the microscopic parameters contained in the Hamiltonian, (Lee and Stone 1985, Al'tshuler 1985). The relation $G_{\mathrm{N}} \gg \delta G$ holds true because the number of propagating channels can be much larger than unity in metallic conductors.

The similar mesoscopic fluctuations exist in the Josephson current in SNS junctions where two spin-singlet s-wave superconductors sandwich a dirty normal metal with its length being L. The current-phase relationship for the jth sample would be given by

$$J^{(j)}(\varphi) = J_1^{(j)} \sin\varphi - J_2^{(j)} \sin 2\varphi. \tag{7.54}$$

When the two superconductors belong to spin-singlet s-wave symmetry class, the relation

$$J_1^{(j)} \gg J_2^{(j)} > 0 \tag{7.55}$$

holds true for all samples. The ensemble average of the Josephson critical current is described well by

$$J_c = \langle J_1 \rangle = \frac{\Delta}{e} G_{\mathrm{N}}. \tag{7.56}$$

The fluctuations of the critical current is given by Al'tshuler and Spivak (1987)

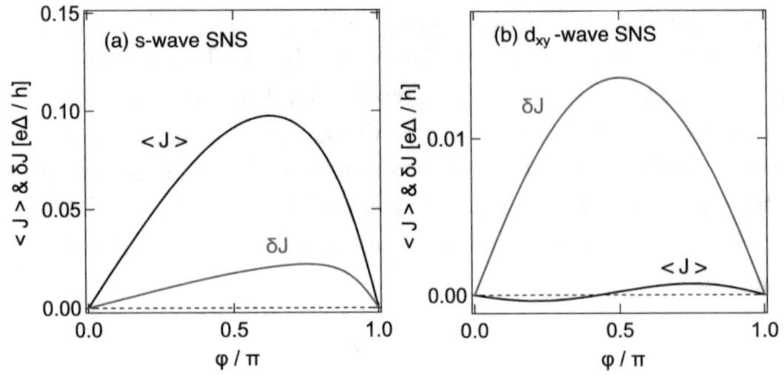

Fig. 7.5 Numerical results by using lattice Green's function method for s-wave SNS junctions in (**a**) and d_{xy}-wave SNS junctions in (**b**). The ensemble average of the Josephson current and its fluctuations are plotted as a function of phase difference across the junction. The number of sample used for the ensemble average is $N_s = 500$ and the temperature is chosen as $T = 0.01T_c$

$$\delta J_c = \frac{eE_{\text{Th}}}{\hbar} C_J, \tag{7.57}$$

for $L \gg \xi_0$, where C_J is a constant of the order of unity depending slightly on the sample geometry. In usual mesoscopic metals, $\delta J_c \ll J_c$ holds true. In experiments, the Josephson current is measured for a single sample with a specific random impurity configuration. The relation $\delta J_c \ll J_c$ suggests that the current amplitudes in a sample $J_1^{(j)}$ would be similar values to J_c. In Fig. 7.5a, we plot the numerical results of the Josephson current in s-wave SNS junctions, where the Josephson current is calculated for 500 samples as a function of φ. The current-phase relationship of $\langle J \rangle$ deviates from the sinusoidal function due to the second harmonic. The numerical results for s-wave junction show $\langle J \rangle < \delta J$ in agreement with the results of the analytic calculations.

Unconventional superconductivity enriches the mesoscopic transport phenomena in SNS junctions. We consider a SNS junction consisting of two spin-singlet d_{xy}-wave superconductor. As discussed in Eq. (7.27), the proximity effect is absent in a normal metal attached to a d_{xy}-wave superconductor. However, this statement is valid only when we consider the pairing correlations after ensemble averaging. In Eq. (7.27), the average over the incident angle of a quasiparticle at the NS interface and the average over different samples play the same role in calculating the pairing correlation. Namely, the absence of the lowest order coupling $\langle J_1 \rangle = 0$ and the relation

$$\langle J(\varphi) \rangle = -\langle J_2 \rangle \, \sin 2\varphi \tag{7.58}$$

are expected in d_{xy}-wave SNS junctions. The results in a numerical simulation agree with the prediction in Eq. (7.58). In Fig. 7.5b, we plot $\langle J \rangle$ and δJ as a function of φ for d_{xy}-wave SNS junctions. The lowest harmonics are absent and the second harmonic is dominant in the average. Simultaneously, the numerical results show the relation

$$\delta J \gg \langle J \rangle \tag{7.59}$$

is satisfied. This relation means that the results after ensemble averaging cannot predict the properties of the Josephson current of a particular sample in an experiment. In Eq. (7.50), $\langle G \rangle$ corresponds to the classical results of the conductance. However, a classical counterpart to the Josephson current is absent because the Josephson current is a result of the phase coherence of an electron. The ensemble-averaged Josephson current is no longer related to the observed current of a single experiment. In numerical simulation, we find that $|J_1^{(j)}| \gg J_2^{(j)} > 0$ holds true for most of the samples. Thus, $\langle J_1 \rangle = 0$ in Fig. 7.5b indicates that half of the samples are of 0-junction and the rest half of the samples are of π-junction. The relation in Eq. (7.59) can be found in various Josephson junctions such as p_y-wave SNS junctions (Asano et al. 2006) and SFS junction (Zyuzin et al. 2003) with F denoting a ferromagnet.

References

Al'tshuler, B.L.: JETP Lett. **41**, 648 (1985)

Al'tshuler, B.L., Spivak, B.Z.: Sov. Phys. JETP **65**, 343 (1987)

Asano, Y.: Phys. Rev. B **64**, 014511 (2001). https://doi.org/10.1103/PhysRevB.64.014511

Asano,Y., Tanaka, Y., Kashiwaya, S.: Phys. Rev. Lett. **96**, 097007 (2006). https://doi.org/10.1103/PhysRevLett.96.097007

Asano, Y., Tanaka, Y., Kashiwaya, S.: Phys. C: Supercond. **460-462**, 1327 (2007). https://doi.org/10.1016/j.physc.2007.04.029. https://www.sciencedirect.com/science/article/pii/S0921453407005369. Proceedings of the 8th International Conference on Materials and Mechanisms of Superconductivity and High Temperature Superconductors

Asano, Y., Tanaka, Y., Yokoyama, T., Kashiwaya, S.: Phys. Rev. B **74**, 064507 (2006). https://doi.org/10.1103/PhysRevB.74.064507

Asano, Y., Golubov, A.A., Fominov, Y.V., Tanaka, Y.: Phys. Rev. Lett. **107**, 087001 (2011). https://doi.org/10.1103/PhysRevLett.107.087001

Berezinskii, V.L.: JETP Lett. **20**, 287 (1974)

Ikegaya, S., Suzuki, S.I., Tanaka, Y., Asano, Y.: Phys. Rev. B **94**, 054512 (2016). https://doi.org/10.1103/PhysRevB.94.054512

Imry, Y.: Introduction to Mesoscopic Physics, 2nd edn. Oxford University Press (2002)

Ishii, C.: Prog. Theor. Phys. **44**(6), 1525 (1970). https://doi.org/10.1143/PTP.44.1525

Kopnin, N.: Theory of Nonequilibrium Superconductivity. Oxford University Press, New York (2001)

Kulik, I.O., Omel'yanchuk, A.N.: Sov. J. Low Temp. Phys. **3**, 945 (1977)

Lee, P.A., Stone, A.D.: Phys. Rev. Lett. **55**, 1622 (1985). https://doi.org/10.1103/PhysRevLett.55.1622

Sato, M., Tanaka, Y., Yada, K., Yokoyama, T.: Phys. Rev. B **83**, 224511 (2011). https://doi.org/10.1103/PhysRevB.83.224511

Suzuki, S.I., Asano, Y.: Phys. Rev. B **91**, 214510 (2015). https://doi.org/10.1103/PhysRevB.91.214510

Tanaka, Y., Golubov, A.A.: Phys. Rev. Lett. **98**, 037003 (2007). https://doi.org/10.1103/PhysRevLett.98.037003

Tanaka, Y., Kashiwaya, S.: Phys. Rev. B **70**, 012507 (2004). https://doi.org/10.1103/PhysRevB.70.012507

Tanaka, Y., Asano, Y., Golubov, A.A., Kashiwaya, S.: Phys. Rev. B **72**, 140503 (2005). https://doi.org/10.1103/PhysRevB.72.140503
Zyuzin, A.Y., Spivak, B., Hruška, M.: Europhys. Lett. (EPL) **62**(1), 97 (2003). https://doi.org/10.1209/epl/i2003-00367-8

Appendix A
Pair Potential Near the Transition Temperature

The dependence of Δ on temperature near T_c can be described analytically. Using an identity,

$$k_B T \sum_{n=-\infty}^{\infty} \frac{1}{\hbar^2\omega_n^2 + a^2} = \frac{1}{2a} \tanh\left(\frac{a}{2k_B T}\right), \quad \omega_n = (2n+1)\pi k_B T/\hbar, \tag{A.1}$$

we first derive a relation from the gap equation,

$$(gN_0)^{-1} = \int_0^{\hbar\omega_D} d\xi \, \frac{1}{E} \tanh\left(\frac{E}{2k_B T}\right) = k_B T \sum_{\omega_n}^{\omega_D} \int_{-\infty}^{\infty} d\xi \, \frac{1}{\hbar^2\omega_n^2 + \xi^2 + \Delta^2}, \tag{A.2}$$

with $E = \sqrt{\xi^2 + \Delta^2}$. Here, we introduce a high-energy cut-off $\hbar\omega_D$ to either the summation or the integration so that the summation and integration converge. At $T = T_c$, by putting $\Delta = 0$, the equation results in

$$(gN_0)^{-1} = k_B T_c \sum_{\omega_n}^{\omega_D} \int_{-\infty}^{\infty} d\xi \, \frac{1}{\omega_n^2 + \xi^2}, \tag{A.3}$$

$$= k_B T_c \sum_{\omega_n}^{\omega_D} \frac{\pi}{|(2n+1)\pi k_B T_c|} = \sum_{n=0}^{N_{T_c}} \frac{1}{n+1/2}, \tag{A.4}$$

$$= \sum_{n=0}^{N_T} \frac{1}{n+1/2} + \sum_{n=0}^{N_{T_c}} \frac{1}{n+1/2} - \sum_{n=0}^{N_T} \frac{1}{n+1/2}, \tag{A.5}$$

$$\approx \sum_{n=0}^{N_T} \frac{1}{n+1/2} + \log\left(\frac{T}{T_c}\right), \tag{A.6}$$

Y. Asano, *Andreev Reflection in Superconducting Junctions*,
SpringerBriefs in Physics, https://doi.org/10.1007/978-981-16-4165-7

where we have used a relation

$$\sum_{n=0}^{N} \frac{1}{n+1/2} \approx \gamma + \log N + 2\log 2 + O\left(\frac{1}{N^2}\right), \tag{A.7}$$

$$N_{T_c} = \left[\frac{\hbar\omega_D}{2\pi k_B T_c}\right]_{\mathrm{G}}, \quad N_T = \left[\frac{\hbar\omega_D}{2\pi k_B T}\right]_{\mathrm{G}}, \tag{A.8}$$

for $N \gg 1$. Here, $\gamma = 0.577215\cdots$ is the Euler constant and $[\cdots]_{\mathrm{G}}$ is the Gauss symbol meaning the integer part of the argument. For $T \lesssim T_c$, Δ is much smaller than T_c. Thus, it is possible to expand the integrand with respect to Δ/ω_n,

$$\begin{aligned}
(gN_0)^{-1} =& 2k_BT \sum_{\omega_n>0}^{\omega_D} \int_{-\infty}^{\infty} d\xi \frac{1}{\hbar^2\omega_n^2 + \xi^2 + \Delta^2}, &\text{(A.9)}\\
\approx& 2k_BT \sum_{\omega_n>0}^{\omega_D} \int_{-\infty}^{\infty} d\xi \left[\frac{1}{\hbar^2\omega_n^2 + \xi^2} - \frac{\Delta^2}{(\hbar^2\omega_n^2 + \xi^2)^2} + \cdots\right], &\text{(A.10)}\\
=& \pi k_BT \sum_{\omega_n>0}^{\omega_D} \left[\frac{1}{\hbar\omega_n} - \frac{1}{(\hbar\omega_n)^3}\frac{\Delta^2}{2}\right], &\text{(A.11)}\\
=& \sum_{n=0}^{N_T} \frac{1}{n+1/2} - \left(\frac{\Delta}{\pi k_BT}\right)^2 \sum_{n=0}^{\infty} \frac{1}{(2n+1)^3}. &\text{(A.12)}
\end{aligned}$$

Using the definition of the ζ-function,

$$\zeta(n) = \sum_{j=1}^{\infty} \frac{1}{j^n}, \quad \sum_{n=0}^{\infty} \frac{1}{(2n+1)^m} = \frac{2^m - 1}{2^m}\zeta(m), \tag{A.13}$$

we reach at a relation of

$$\log\left(\frac{T}{T_c}\right) = -\left(\frac{\Delta}{\pi k_BT}\right)^2 \frac{7\zeta(3)}{8}. \tag{A.14}$$

Finally, we obtain an expression of the pair potential near the transition temperature,

$$\Delta = \pi k_B T_c \sqrt{\frac{8}{7\zeta(3)}} \sqrt{\frac{T_c - T}{T_c}}. \tag{A.15}$$

Appendix B
Conductance Formula and Transport Channels

Landauer's conductance formula can be derived from an intuitive argument. To measure the conductance, we consider one-dimensional junction as shown in Fig. B.1, where a sample is connected to two perfect read wires which are terminated by reservoirs. The perfect lead wire is free from any scattering events. The reservoir is always in equilibrium characterized by a chemical potential and absorbs any incoming waves. In Fig. B.1, the chemical potential on the left (right) reservoir is μ_L (μ_R). The sample is shorter than any inelastic scattering lengths. An electron at the left reservoir can go into the sample and reach the right reservoir. Since all the states with energy below μ_R are occupied, only an electron at the energy window $\mu_L - \mu_R = eV \ll \mu_R$ can penetrate into the left lead wire and the sample. The electron number in such an energy window is estimated as

$$N = \frac{eV}{\pi \hbar v_F}, \tag{B.1}$$

where $(\pi \hbar v_F)^{-1}$ is the density of states per unit length per spin in one-dimensional conductor. Among them, an electron with the positive velocity v_F penetrates into the sample

$$\frac{N}{2} v_F T = \frac{eV}{\pi \hbar v_F} v_F \frac{1}{2} T, \tag{B.2}$$

and it can transmit to the right reservoir, where T denotes the transmission probability of the sample. The electric current is calculated as

$$J = e \frac{N}{2} v_F T = \frac{e^2}{h} T V. \tag{B.3}$$

The conductance defined by the relation $J = GV$ results in

Y. Asano, *Andreev Reflection in Superconducting Junctions*,
SpringerBriefs in Physics, https://doi.org/10.1007/978-981-16-4165-7

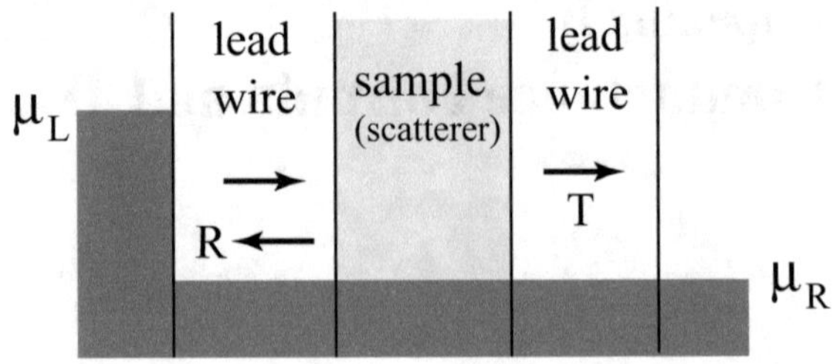

Fig. B.1 A Schematic picture of a junction considered in Landauer's conductance formula

$$G = \frac{e^2}{h} T. \tag{B.4}$$

The transmission probability T and the reflection probability R are the characteristic physical values of waves. The current conservation law implies $T + R = 1$. The interference effect of an electron is taken into account through these transport probabilities. The product of the velocity and the density of states gives a constant, which is a characteristic feature in one-dimensional conductors. Thus, the formula can be applied to any quasi-one-dimensional structures which have more than one transport channel in lead wires. The formula for such case is given by

$$G = \frac{e^2}{h} \sum_m T_m, \tag{B.5}$$

where T_m is the transmission probability of an electron which is incoming to the sample from the mth propagating channel as shown in Fig. B.2.

The perfect lead wires are needed to define the transport coefficients. To see this, let us consider a quasi-one-dimensional conductor. The Schrödinger equation is given by

$$\left[-\frac{\hbar^2 \nabla^2}{2m} - \epsilon_F + V(\boldsymbol{r}) \right] \phi(x, y) = E\phi(x, y), \tag{B.6}$$

where $V(\boldsymbol{r})$ represents scattering potential in a sample and confining potential of a quasi-one-dimensional junction. We apply the periodic boundary condition in the transverse direction to the current (y-direction), which implies a conductor is confined within a finite region W in the y-direction. As shown in Fig. B.2a, the sample is connected to two lead wires. To define the transport channels clearly, we assume that the lead wires are free from any scatterers. The solution of the Schrödinger equation at the lead wire can described as

$$\phi(x, y) = e^{ikx} f_p(y), \quad f_p(y) = \frac{e^{ipy}}{\sqrt{W}}, \tag{B.7}$$

$$E_{k,p_m} = \frac{\hbar^2}{2m}\left(k^2 + p_m^2\right), \quad p_m = \frac{2\pi m}{W}, \quad m = 0, \pm 1, \pm 2, \ldots. \tag{B.8}$$

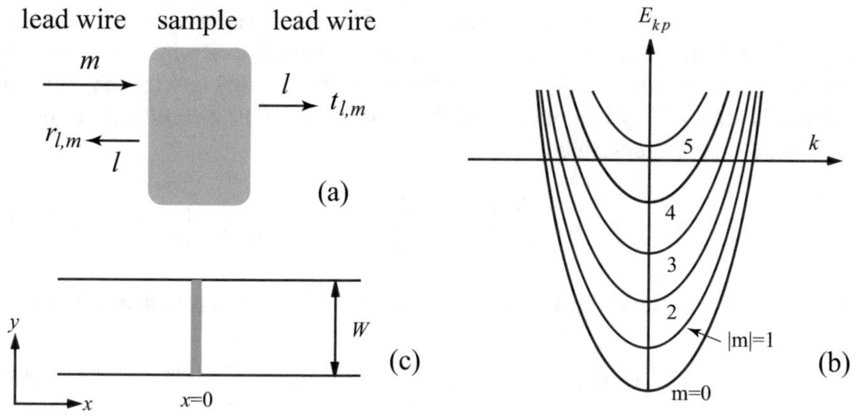

Fig. B.2 **a** The transport channels can be defined in perfect lead wires, where l and m indicate the propagating channels. **b** The subband structures in a quasi-one-dimensional lead wire. **c** Two lead wires are separated by a potential barrier at $x = 0$, where the width of the junction is W

In Fig. B.2b, we plot the energy $E_{k,p}$ as a function of the wavenumber in the x-direction k. The kinetic energy in the y-direction is quantized due to the confinement potential as shown with $m = 0$, $m = \pm 1$..... Thus, the dispersion shows the subband structures, and m specifies the subband. The horizontal line indicates the Fermi level. At the Fermi level, the relation $E_{k,p} = \epsilon_F$ defines the wavenumber in the x-direction

$$k_m^2 = \frac{2m\epsilon_F}{\hbar^2} - p_m^2. \tag{B.9}$$

When k_m is real, the wave function is a plane wave with a group velocity of $v_l = \hbar k_m/m$. Such a transport channel is called a propagating channel and can carry the current. When k_m is imaginary, on the other hand, the wave function decays exponentially in the x-direction. Such a channel is called an evanescent channel and cannot carry the current. In this picture, nine subbands indicated by $m = 0, \pm 1, \pm 2, \pm 3, \pm 4$ are propagating and the subbands by $|m| \geq 5$ are evanescent. The transport coefficients are defined between the propagating channels.

Next, we consider a situation where an electron is incident on the left lead from the mth channel. The incident electron either transmits to the right lead wire or is reflected to the left lead wire. The wave functions can be described as

$$\phi_m^L(x, y) = e^{ik_m x} f_m(y) + \sum_l r_{l,m} e^{-ik_l x} f_l(y), \tag{B.10}$$

$$\phi_m^R(x, y) = \sum_l t_{l,m} e^{ik_l x} f_l(y), \tag{B.11}$$

where $t_{l,m}$ is the transmission coefficient to the lth channel at the right lead wire and $r_{l,m}$ is the reflection coefficient to the lth channel at the left lead wire. The outgoing channel l is not necessarily identical to the incoming channel m when an incident electron is scattered by random potentials at the sample. The electric current flows in the lead wires are described as

$$J_j = \int_0^W dy \frac{e\hbar}{2mi} \left[(\phi_m^j)^* \frac{d}{dx} \phi_m^j - \frac{d}{dx} (\phi_m^j)^* \phi_m^j \right], \tag{B.12}$$

for $j = L$ and R. The current conservation law implies $J_L = J_R$, which leads to

$$\sum_l |t_{l,m}|^2 \frac{v_l}{v_m} + |r_{l,m}|^2 \frac{v_l}{v_m} = 1, \quad v_l = \frac{\hbar k_l}{m}, \tag{B.13}$$

where we have used the orthonormality of the wave function

$$\int_0^W dy\ f_n^*(y) f_l(y) = \delta_{n,l}. \tag{B.14}$$

These transport coefficients are calculated from the boundary condition of the wave function. In Fig. B.2c, we consider a situation where the sample is represented by the potential barrier of $v_0 \delta(x)$. A boundary condition for the wave functions,

$$\phi_m^L(0, y) = \phi_m^R(0, y), \tag{B.15}$$

is derived because the wave function must be single-valued. By substituting Eqs. (B.10) and (B.11) into the boundary condition and carrying out the integration along y after multiplying $f_n^*(y)$, we find

$$\delta_{n,m} + r_{n,m} = t_{n,m}. \tag{B.16}$$

To derive the second boundary condition, we integrate the Schrödinger equation at an short interval of $-\gamma < x < \gamma$ and take the limit of $\gamma \to 0$,

$$\lim_{\gamma \to 0} \int_{-\gamma}^{\gamma} dx \left[-\frac{\hbar^2}{2m} \left(\frac{\partial^2}{\partial x^2} + \frac{\partial^2}{\partial y^2} \right) - \epsilon_F + v_0 \delta(x) \right] \phi(x, y)$$
$$= \lim_{\gamma \to 0} \int_{-\gamma}^{\gamma} dx\ E\phi(x, y). \tag{B.17}$$

Since the right-hand side becomes zero, we obtain

$$-\frac{\hbar^2}{2m} \left[\frac{\partial}{\partial x} \phi^R(x, y) \bigg|_{x=0} - \frac{\partial}{\partial x} \phi^L(x, y) \bigg|_{x=0} \right] + v_0 \phi^R(0, y) = 0. \tag{B.18}$$

By substituting Eqs. (B.10) and (B.11) into the boundary condition and carrying out the integration along y after multiplying $f_n^*(y)$, we find the relation

$$\bar{k}_m \, \delta_{n,m} - \bar{k}_n \, r_{n,m} = (\bar{k}_n + 2i z_0) t_{n,m}, \tag{B.19}$$

$$z_0 = \frac{m \, v_0}{\hbar^2 k_F}, \quad \bar{k}_m = \frac{k_m}{k_F}, \quad \frac{\hbar^2 k_F^2}{2m} = \epsilon_F. \tag{B.20}$$

From (B.16) and (B.19), the transport coefficients are calculated as

$$t_{n,m} = \frac{\bar{k}_m}{\bar{k}_m + i z_0} \delta_{n,m}, \quad r_{n,m} = \frac{-i z_0}{\bar{k}_m + i z_0} \delta_{n,m}. \tag{B.21}$$

Since the barrier potential retains translational symmetry in the y-direction, the transport coefficient is diagonal with respect to the channel index. The conductance of the potential barrier results in

$$G = \frac{2e^2}{h} \mathrm{Tr} \left[\hat{t} \, \hat{t}^\dagger \right] = \frac{2e^2}{h} \sum_m |t_{m,m}|^2, \tag{B.22}$$

where we multiply 2 to the conductance by taking the spin degeneracy into account. For $W k_F \gg 1$, the summation over m can be replaced by the integral for wavenumber in the y-direction ,

$$\sum_m = \frac{W}{2\pi} \int_{-k_F}^{k_F} dp = \left[\frac{W k_F}{\pi} \right]_{\mathrm{G}} = N_c, \tag{B.23}$$

where $[\cdots]_{\mathrm{G}}$ is the Gauss symbol and N_c represents the number of propagating channels on the Fermi surface. When the potential barrier is absent, the transmission probability is unity for all propagating channels. The conductance of such a perfect conductor remains finite value of

$$G = \frac{2e^2}{h} N_c, \tag{B.24}$$

because the width of the junction is finite. The conductance in Eq. (B.24) is called the Sharvin conductance in mesoscopic physics.

Normal transport coefficients through a magnetic potential barrier

We consider the transmission of an electron through a magnetically active insulating barrier. The Hamiltonian reads

$$\hat{H}_{\mathrm{N}}(\boldsymbol{r}) = \left\{ -\frac{\hbar^2 \nabla^2}{2m} - \epsilon_{\mathrm{F}} + V_0 \delta(x) \right\} \hat{\sigma}_0 + \boldsymbol{V} \cdot \hat{\boldsymbol{\sigma}} \delta(x), \tag{B.25}$$

where $\hat{\sigma}_j$ for $j = 1 - 3$ is Pauli's matrix in spin space. The wave function of an electron on either side of the interface can be described as

$$\psi_L(\boldsymbol{r}) = \left[\boldsymbol{a}e^{ik_x x} + \boldsymbol{A}e^{-ik_x x}\right] e^{i\boldsymbol{p}\cdot\boldsymbol{\rho}}, \tag{B.26}$$

$$\psi_R(\boldsymbol{r}) = \left[\boldsymbol{b}e^{-ik_x x} + \boldsymbol{B}e^{ik_x x}\right] e^{i\boldsymbol{p}\cdot\boldsymbol{\rho}}, \tag{B.27}$$

$$\boldsymbol{a} = \begin{bmatrix} \alpha_\uparrow \\ \alpha_\downarrow \end{bmatrix}, \ \boldsymbol{b} = \begin{bmatrix} \beta_\uparrow \\ \beta_\downarrow \end{bmatrix}, \ \boldsymbol{A} = \begin{bmatrix} A_\uparrow \\ A_\downarrow \end{bmatrix}, \ \boldsymbol{B} = \begin{bmatrix} B_\uparrow \\ B_\downarrow \end{bmatrix}, \tag{B.28}$$

where $\boldsymbol{r} = (x, \boldsymbol{\rho})$, $k_x^2 + \boldsymbol{p}^2 = k_F^2 = 2m\epsilon_{\mathrm{F}}/\hbar^2$, and $\boldsymbol{a}$ and $\boldsymbol{b}$ represent the amplitudes of incoming wave into the interface. On the other hand, $\boldsymbol{A}$ and $\boldsymbol{B}$ represent the amplitudes of outgoing wave from the interface. The two wave functions are connected by the the boundary conditions at $x = 0$,

$$\psi_L(0, \boldsymbol{\rho}) = \psi_R(0, \boldsymbol{\rho}), \tag{B.29}$$

$$-\frac{\hbar^2}{2m}\left[\frac{\partial}{\partial x}\psi_R(\boldsymbol{r}) - \frac{\partial}{\partial x}\psi_L(\boldsymbol{r})\right]_{x=0} + \left(V_0 + \boldsymbol{V}\cdot\hat{\boldsymbol{\sigma}}\right)\psi_L(0, \boldsymbol{\rho}) = 0. \tag{B.30}$$

The boundary conditions relate the outgoing waves and incoming ones as

$$\begin{bmatrix} \boldsymbol{A} \\ \boldsymbol{B} \end{bmatrix} = \begin{bmatrix} \hat{r}_{\mathrm{N}} & \hat{t}'_{\mathrm{N}} \\ \hat{t}_{\mathrm{N}} & \hat{r}'_{\mathrm{N}} \end{bmatrix} \begin{bmatrix} \boldsymbol{a} \\ \boldsymbol{b} \end{bmatrix}, \tag{B.31}$$

$$\hat{t}_{\mathrm{N}} = \hat{t}'_{\mathrm{N}} = \frac{1}{\Xi} k \left[k + iz_0 - i\boldsymbol{z}\cdot\hat{\boldsymbol{\sigma}}\right], \tag{B.32}$$

$$\hat{r}_{\mathrm{N}} = \hat{r}'_{\mathrm{N}} = \frac{1}{\Xi}\left[(k + iz_0)(-iz_0) - |\boldsymbol{z}|^2 - k\, i\boldsymbol{z}\cdot\hat{\boldsymbol{\sigma}}\right], \tag{B.33}$$

$$\boldsymbol{z} = \frac{m\boldsymbol{V}}{\hbar^2 k_F}, \quad \Xi = (k + iz_0)^2 + |\boldsymbol{z}|^2. \tag{B.34}$$

The current conservation law implies

$$\hat{t}_{\mathrm{N}}\hat{t}_{\mathrm{N}}^\dagger + \hat{r}_{\mathrm{N}}\hat{r}_{\mathrm{N}}^\dagger = \hat{\sigma}_0. \tag{B.35}$$

The transport coefficients in the hole branch of BdG picture are calculated as

$$\underline{\hat{t}_{\mathrm{N}}} = \frac{1}{\Xi^*} k\left[k - iz_0 + i\boldsymbol{z}\cdot\hat{\boldsymbol{\sigma}}^*\right] = \hat{t}_{\mathrm{N}}^*, \tag{B.36}$$

$$\underline{\hat{r}_{\mathrm{N}}} = \frac{1}{\Xi^*}\left[(k - iz_0)(iz_0) - |\boldsymbol{z}|^2 + k\, i\boldsymbol{z}\cdot\hat{\boldsymbol{\sigma}}^*\right] = \hat{r}_{\mathrm{N}}^*. \tag{B.37}$$

Appendix C
Mean-Field Theory in Real Space

The Hamiltonian for interacting electrons is represented by

$$\mathcal{H} = \mathcal{H}_{\mathrm{N}} + \mathcal{H}_{\mathrm{I}}, \tag{C.1}$$

$$\mathcal{H}_{\mathrm{N}} = \int d\boldsymbol{r} \sum_{\alpha,\beta} \psi_{\alpha}^{\dagger}(\boldsymbol{r}) \xi_{\alpha,\beta}(\boldsymbol{r}) \psi_{\alpha}(\boldsymbol{r}), \tag{C.2}$$

$$\hat{\xi}_{\alpha,\beta}(\boldsymbol{r}) = \left\{ -\frac{\hbar^2}{2m}\left(\nabla - \frac{ie}{\hbar c}\boldsymbol{A}\right)^2 - \epsilon_F + V_0(\boldsymbol{r}) \right\} \hat{\sigma}_0 + \boldsymbol{V}(\boldsymbol{r}) \cdot \hat{\boldsymbol{\sigma}}$$
$$+ i\boldsymbol{\lambda}(\boldsymbol{r}) \times \nabla \cdot \hat{\boldsymbol{\sigma}}, \tag{C.3}$$

$$\mathcal{H}_{\mathrm{I}} = \frac{1}{2} \int d\boldsymbol{r} \int d\boldsymbol{r}' \sum_{\alpha,\beta} \psi_{\beta}^{\dagger}(\boldsymbol{r}')\, \psi_{\alpha}^{\dagger}(\boldsymbol{r}) \left[-g(\boldsymbol{r} - \boldsymbol{r}') \right] \psi_{\alpha}(\boldsymbol{r})\, \psi_{\beta}(\boldsymbol{r}'), \tag{C.4}$$

where V_0 is the spin-independent potential, $\boldsymbol{V}$ is the Zeeman potential, $\boldsymbol{\lambda}$ represents the spin-orbit potential, and $g(\boldsymbol{r} - \boldsymbol{r}') = g(\boldsymbol{r}' - \boldsymbol{r}) > 0$ is the attractive interaction between two electrons. The pair potential is defined as

$$\Delta_{\alpha,\beta}(\boldsymbol{r}, \boldsymbol{r}') = -\, g(\boldsymbol{r} - \boldsymbol{r}') \left\langle \psi_{\alpha}(\boldsymbol{r})\, \psi_{\beta}(\boldsymbol{r}') \right\rangle, \tag{C.5}$$

$$\Delta_{\alpha,\beta}^{*}(\boldsymbol{r}, \boldsymbol{r}') = -\, g(\boldsymbol{r} - \boldsymbol{r}') \left\langle \psi_{\beta}^{\dagger}(\boldsymbol{r}')\, \psi_{\alpha}^{\dagger}(\boldsymbol{r}) \right\rangle = g(\boldsymbol{r} - \boldsymbol{r}') \left\langle \psi_{\alpha}^{\dagger}(\boldsymbol{r})\, \psi_{\beta}^{\dagger}(\boldsymbol{r}') \right\rangle. \tag{C.6}$$

The pair potential satisfies the antisymmetric property

$$\Delta_{\alpha,\beta}(\boldsymbol{r}, \boldsymbol{r}') = -\Delta_{\beta,\alpha}(\boldsymbol{r}', \boldsymbol{r}), \tag{C.7}$$

because of the anticommutation relation among the electron operators. The interaction Hamiltonian is decoupled by the mean-field approximation. The operators are replaced by the order parameters as

Y. Asano, *Andreev Reflection in Superconducting Junctions*,
SpringerBriefs in Physics, https://doi.org/10.1007/978-981-16-4165-7

$$\psi_\beta^\dagger(\boldsymbol{r}')\,\psi_\alpha^\dagger(\boldsymbol{r}) = -\frac{\Delta_{\alpha,\beta}^*(\boldsymbol{r},\boldsymbol{r}')}{g(\boldsymbol{r}-\boldsymbol{r}')} + \left[\psi_\beta^\dagger(\boldsymbol{r}')\,\psi_\alpha^\dagger(\boldsymbol{r}) + \frac{\Delta_{\alpha,\beta}^*(\boldsymbol{r},\boldsymbol{r}')}{g(\boldsymbol{r}-\boldsymbol{r}')}\right], \tag{C.8}$$

$$\psi_\alpha(\boldsymbol{r})\,\psi_\beta(\boldsymbol{r}') = -\frac{\Delta_{\alpha,\beta}(\boldsymbol{r},\boldsymbol{r}')}{g(\boldsymbol{r}-\boldsymbol{r}')} + \left[\psi_\alpha(\boldsymbol{r})\,\psi_\beta(\boldsymbol{r}') + \frac{\Delta_{\alpha,\beta}(\boldsymbol{r},\boldsymbol{r}')}{g(\boldsymbol{r}-\boldsymbol{r}')}\right]. \tag{C.9}$$

The first term is the average of the operator and the second term represents the fluctuations. The mean-field expansion is carried out up to the first order of the fluctuations,

$$\mathcal{H}_\mathrm{I} \to \frac{1}{2}\int d\boldsymbol{r}\int d\boldsymbol{r}' \sum_{\alpha,\beta} \psi_\alpha^\dagger(\boldsymbol{r})\Delta_{\alpha,\beta}(\boldsymbol{r},\boldsymbol{r}')\,\psi_\beta^\dagger(\boldsymbol{r}') - \psi_\alpha(\boldsymbol{r})\,\Delta_{\alpha,\beta}^*(\boldsymbol{r},\boldsymbol{r}')\,\psi_\beta(\boldsymbol{r}') - \frac{|\Delta_{\alpha,\beta}(\boldsymbol{r},\boldsymbol{r}')|^2}{g(\boldsymbol{r}-\boldsymbol{r}')}. \tag{C.10}$$

As a result, the mean-field Hamiltonian of superconductivity is represented as

$$\mathcal{H}_\mathrm{MF} = \frac{1}{2}\int d\boldsymbol{r}\int d\boldsymbol{r}' \sum_{\alpha,\beta} \left[\psi_\uparrow^\dagger(\boldsymbol{r}),\,\psi_\downarrow^\dagger(\boldsymbol{r}),\,\psi_\uparrow(\boldsymbol{r}),\,\psi_\downarrow(\boldsymbol{r})\right] \times \begin{bmatrix} \delta(\boldsymbol{r}-\boldsymbol{r}')\,\hat{\xi}(\boldsymbol{r}') & \hat{\Delta}(\boldsymbol{r},\boldsymbol{r}') \\ -\hat{\Delta}^*(\boldsymbol{r},\boldsymbol{r}') & -\delta(\boldsymbol{r}-\boldsymbol{r}')\,\hat{\xi}^*(\boldsymbol{r}') \end{bmatrix} \begin{bmatrix} \psi_\uparrow(\boldsymbol{r}') \\ \psi_\downarrow(\boldsymbol{r}') \\ \psi_\uparrow^\dagger(\boldsymbol{r}') \\ \psi_\downarrow^\dagger(\boldsymbol{r}') \end{bmatrix}. \tag{C.11}$$

The BdG equation reads

$$\int d\boldsymbol{r}' \sum_{\alpha,\beta} \begin{bmatrix} \delta(\boldsymbol{r}-\boldsymbol{r}')\,\hat{\xi}(\boldsymbol{r}') & \hat{\Delta}(\boldsymbol{r},\boldsymbol{r}') \\ -\hat{\Delta}^*(\boldsymbol{r},\boldsymbol{r}') & -\delta(\boldsymbol{r}-\boldsymbol{r}')\,\hat{\xi}^*(\boldsymbol{r}') \end{bmatrix} \begin{bmatrix} \hat{u}_\nu(\boldsymbol{r}') \\ \hat{v}_\nu(\boldsymbol{r}') \end{bmatrix} = \begin{bmatrix} \hat{u}_\nu(\boldsymbol{r}) \\ \hat{v}_\nu(\boldsymbol{r}) \end{bmatrix} \hat{E}_\nu, \tag{C.12}$$

$$\hat{E}_\nu = \begin{bmatrix} E_{\nu,1} & 0 \\ 0 & E_{\nu,2} \end{bmatrix}, \quad \begin{bmatrix} \hat{u}_\nu(\boldsymbol{r}) \\ \hat{v}_\nu(\boldsymbol{r}) \end{bmatrix} = \begin{bmatrix} u_{\nu,\uparrow,1}(\boldsymbol{r}) & u_{\nu,\uparrow,2}(\boldsymbol{r}) \\ u_{\nu,\downarrow,1}(\boldsymbol{r}) & u_{\nu,\downarrow,2}(\boldsymbol{r}) \\ v_{\nu,\uparrow,1}(\boldsymbol{r}) & v_{\nu,\uparrow,2}(\boldsymbol{r}) \\ v_{\nu,\downarrow,1}(\boldsymbol{r}) & v_{\nu,\downarrow,2}(\boldsymbol{r}) \end{bmatrix}. \tag{C.13}$$

The wave function

$$\begin{bmatrix} \hat{v}_\nu^*(\boldsymbol{r}) \\ \hat{u}_\nu^*(\boldsymbol{r}) \end{bmatrix} \tag{C.14}$$

belongs to the eigenvalue of $-\hat{E}_\nu$. The wave function satisfies the orthonormality and the completeness,

$$\int d\boldsymbol{r} \left[\hat{u}_\lambda^\dagger(\boldsymbol{r}), \hat{v}_\lambda^\dagger(\boldsymbol{r})\right] \begin{bmatrix} \hat{u}_\nu(\boldsymbol{r}) \\ \hat{v}_\nu(\boldsymbol{r}) \end{bmatrix} = \delta_{\lambda,\nu}, \tag{C.15}$$

$$\int d\boldsymbol{r} \left[\hat{u}_\lambda^\dagger(\boldsymbol{r}), \hat{v}_\lambda^\dagger(\boldsymbol{r})\right] \begin{bmatrix} \hat{v}_\nu^*(\boldsymbol{r}) \\ \hat{u}_\nu^*(\boldsymbol{r}) \end{bmatrix} = 0, \tag{C.16}$$

$$\sum_\nu \begin{bmatrix} \hat{u}_\nu(\boldsymbol{r}) \\ \hat{v}_\nu(\boldsymbol{r}) \end{bmatrix} \left[\hat{u}_\nu^\dagger(\boldsymbol{r}'), \hat{v}_\nu^\dagger(\boldsymbol{r}')\right] + \begin{bmatrix} \hat{v}_\nu^*(\boldsymbol{r}) \\ \hat{u}_\nu^*(\boldsymbol{r}) \end{bmatrix} \left[\hat{v}_\nu^{\mathrm{T}}(\boldsymbol{r}'), \hat{u}_\nu^{\mathrm{T}}(\boldsymbol{r}')\right] = \check{1}_{4\times 4}\, \delta(\boldsymbol{r} - \boldsymbol{r}'). \tag{C.17}$$

Let us assume that $\boldsymbol{V}, \boldsymbol{\lambda}$ are uniform, and $\boldsymbol{A} = V_0 = 0$. It is possible to derive the mean-field Hamiltonian in momentum space by substitute the Fourier transformation

$$\psi_\alpha(\boldsymbol{r}) = \frac{1}{\sqrt{V_{\mathrm{vol}}}} \sum_{\boldsymbol{k}} \psi_{\boldsymbol{k},\alpha} e^{i\boldsymbol{k}\cdot\boldsymbol{r}}, \tag{C.18}$$

$$\Delta_{\alpha,\beta}(\boldsymbol{r} - \boldsymbol{r}') = \frac{1}{V_{\mathrm{vol}}} \sum_{\boldsymbol{q}} \Delta_{\alpha,\beta}(\boldsymbol{q}) e^{i\boldsymbol{k}\cdot(\boldsymbol{r}-\boldsymbol{r}')}, \tag{C.19}$$

into Eq. (C.11),

$$\begin{aligned}\mathcal{H}_{\mathrm{MF}} =& \frac{1}{2}\int d\boldsymbol{r} \int d\boldsymbol{r}' \sum_{\alpha,\beta} \frac{1}{V_{\mathrm{vol}}} \sum_{\boldsymbol{k},\boldsymbol{k}'} \left[\psi_{\boldsymbol{k}',\uparrow}^\dagger, \psi_{\boldsymbol{k}',\downarrow}^\dagger, \psi_{-\boldsymbol{k}',\uparrow}, \psi_{-\boldsymbol{k}',\downarrow}\right] e^{-i\boldsymbol{k}'\cdot\boldsymbol{r}} \\ &\times \begin{bmatrix} \delta(\boldsymbol{r}-\boldsymbol{r}')\,\hat{\xi}(\boldsymbol{r}') & \frac{1}{V_{\mathrm{vol}}}\sum_{\boldsymbol{q}} \hat{\Delta}(\boldsymbol{q}) e^{i\boldsymbol{q}\cdot(\boldsymbol{r}-\boldsymbol{r}')} \\ -\frac{1}{V_{\mathrm{vol}}}\sum_{\boldsymbol{q}} \hat{\Delta}^*(\boldsymbol{q}) e^{-i\boldsymbol{q}\cdot(\boldsymbol{r}-\boldsymbol{r}')} & -\delta(\boldsymbol{r}-\boldsymbol{r}')\,\hat{\xi}^*(\boldsymbol{r}') \end{bmatrix} e^{i\boldsymbol{k}\cdot\boldsymbol{r}'} \begin{bmatrix} \psi_{\boldsymbol{k},\uparrow} \\ \psi_{\boldsymbol{k},\downarrow} \\ \psi_{-\boldsymbol{k},\uparrow}^\dagger \\ \psi_{-\boldsymbol{k},\downarrow}^\dagger \end{bmatrix},\end{aligned} \tag{C.20}$$

$$\begin{aligned} =& \frac{1}{2}\int d\boldsymbol{r} \sum_{\alpha,\beta} \frac{1}{V_{\mathrm{vol}}} \sum_{\boldsymbol{k},\boldsymbol{k}'} \left[\psi_{\boldsymbol{k}',\uparrow}^\dagger, \psi_{\boldsymbol{k}',\downarrow}^\dagger, \psi_{-\boldsymbol{k}',\uparrow}, \psi_{-\boldsymbol{k}',\downarrow}\right] e^{-i\boldsymbol{k}'\cdot\boldsymbol{r}} \\ &\times \begin{bmatrix} \hat{\xi}(\boldsymbol{r}) & \hat{\Delta}(\boldsymbol{k}) \\ -\hat{\Delta}^*(-\boldsymbol{k}) & -\hat{\xi}^*(\boldsymbol{r}) \end{bmatrix} e^{i\boldsymbol{k}\cdot\boldsymbol{r}} \begin{bmatrix} \psi_{\boldsymbol{k},\uparrow} \\ \psi_{\boldsymbol{k},\downarrow} \\ \psi_{-\boldsymbol{k},\uparrow}^\dagger \\ \psi_{-\boldsymbol{k},\downarrow}^\dagger \end{bmatrix},\end{aligned} \tag{C.21}$$

$$
= \frac{1}{2} \sum_{\alpha,\beta} \sum_{k} \left[\psi_{k,\uparrow}, \psi_{k,\downarrow}^{\dagger}, \psi_{-k,\uparrow}, \psi_{-k,\downarrow} \right] \begin{bmatrix} \hat{\xi}_{k} & \hat{\Delta}(k) \\ -\hat{\Delta}^{*}(-k) & -\hat{\xi}_{-k}^{*} \end{bmatrix} \begin{bmatrix} \psi_{k,\uparrow} \\ \psi_{k,\downarrow} \\ \psi_{-k,\uparrow}^{\dagger} \\ \psi_{-k,\downarrow}^{\dagger} \end{bmatrix}, \tag{C.22}
$$

$$
\hat{\xi}_{k} = \xi_{k}\hat{\sigma}_0 + \boldsymbol{V} \cdot \hat{\boldsymbol{\sigma}} - \boldsymbol{\lambda} \times \boldsymbol{k} \cdot \hat{\boldsymbol{\sigma}}, \quad \xi_{k} = \frac{\hbar^2 \boldsymbol{k}^2}{2m} - \epsilon_F. \tag{C.23}
$$

The examples of the pair potential are shown below.

$$
\hat{\Delta}(\boldsymbol{k}) = \begin{cases} \Delta i\hat{\sigma}_2 & : s\text{-wave} \\ \Delta(\bar{k}_x^2 - \bar{k}_y^2) i\hat{\sigma}_2 & : d_{x^2-y^2}\text{-wave} \\ \Delta 2\bar{k}_x \bar{k}_y i\hat{\sigma}_2 & : d_{xy}\text{-wave} \end{cases}, \tag{C.24}
$$

for spin-singlet even-parity order with $\bar{k}_j = k_j/k_F$ for $j = x, y, z$ and

$$
\hat{\Delta}(\boldsymbol{k}) = \begin{cases} \Delta\bar{k}_x\, i\boldsymbol{d} \cdot \hat{\boldsymbol{\sigma}}\, \hat{\sigma}_2 & : p_x\text{-wave} \\ \Delta(\bar{k}_y\hat{\sigma}_1 - \bar{k}_x\hat{\sigma}_2)\, i\hat{\sigma}_2 & : \text{2D Helical } p\text{-wave} \\ \Delta(\bar{k}_x + i\bar{k}_y)\boldsymbol{d} \cdot \hat{\boldsymbol{\sigma}}\, i\hat{\sigma}_2 & : \text{2D chiral } p\text{-wave} \\ \Delta(\bar{k}_x\hat{\sigma}_1 + \bar{k}_y\hat{\sigma}_2 + \bar{k}_z\hat{\sigma}_3)\, i\hat{\sigma}_2 & : \text{Superfluid } {}^3\text{He B phase} \end{cases}, \tag{C.25}
$$

for spin-triplet odd-parity order.

Index

Y. Asano, *Andreev Reflection in Superconducting Junctions*,
SpringerBriefs in Physics, https://doi.org/10.1007/978-981-16-4165-7

Printed in the United States
by Baker & Taylor Publisher Services